LOOKING AT ANIMALS

LOOKING AT
ANIMALS

A Zoologist in Africa

HUGH B. COTT
Sc.D., D.Sc., F.R.P.S.

COLLINS
ST JAMES'S PLACE, LONDON

William Collins Sons & Co Ltd
London · Glasgow · Sydney · Auckland
Toronto · Johannesburg

First published 1975
© Hugh B. Cott 1975
ISBN 0 00 219093 1

Printed and bound in Spain by
E. Belgas, S. L. Ntra. Sra. de la Cabeza, 2 - Bilbao. Spain

Contents

CONTENTS

Photographs

7

PHOTOGRAPHS

PHOTOGRAPHS

All photographs and drawings in this book are by the author.

Preface

The pleasures of travel – diverse as they are – fall broadly into three categories, those of anticipation, realization, and retrospection. The first kind could not be better summarized than in the words of Sir Richard Burton when he wrote: 'The gladdest moment in human life, methinks, is the departure upon a distant journey. Shaking off with one mighty effort the fetters of habit, the leaden weight of routine, the cloak of many cares man feels once more happy . . . A journey, in fact, appeals to imagination, to memory, to hope – the sister Graces of our mortal being.' Encounters with wild creatures in a tropical African setting leave one in no doubt that here we are indeed looking at one of the world's greatest wonders – a serene spectacle of the past. The reality far exceeds expectation.

It has been my privilege, since the inception of Swans Big Game and Birds Safaris ten years ago, to accompany many of the tours as Guest Lecturer. This book has grown out of notes which formed a basis for informal talks given in the field to visitors. My aim has been to provide, however incompletely, an account of the natural history of the more common, or more spectacular or interesting animals likely to be met with in the National Parks and game reserves of Kenya, Tanzania and Uganda. I have drawn attention to the essential unity of life; to interspecific relationships both of competition and co-operation; and to adaptations of structure and behaviour that serve animals variously in their different methods of safety-seeking, food-getting, and reproduction.

I venture to hope the book will whet the appetite of the intending traveller, deepen his appreciation of events observed in the field, and – especially through the medium of photographs and drawings – vividly recall memories of wild places and their incomparable plant and animal life. If in addition the work gives some lasting pleasure to others who cannot make the journey a first-hand experience, I shall be well content.

With recent political changes, some older (and through long-usage more familiar) topographical names have been replaced. I have used the current terms, and the reader is reminded that: Murchison Falls (so named by Sir Samuel Baker in honour of the President of the Royal Geographical Society) and the Park of the same name are now known as Kabalega Falls; Queen Elizabeth Park has become Rwenzori National Park; and Lake Albert and Lake Edward are renamed, respectively, Lake Mobutu, and Lake Idi Amin Dada.

Acknowledgements

The field work done since 1952 could not have been undertaken without practical help generously given by many people. Amongst others too numerous to name here individually I particularly wish to thank R. S. A. Beauchamp, Roger Wheater, John Savidge, Keith Eltringham and Chiels Margach, all of whom gave hospitality, encouragement and help in Uganda. On various excursions I was accompanied by my wife, who shared experiences and excitements of camp life at Magungu and Fajao below Kabalega Falls in 1956, and who lightened my work in many ways. My warm thanks are also due to Ranger Justian Ogwal, for his unfailing assistance during four season's research on the Victoria Nile.

R. Murray Watson, when studying wildebeest in Tanzania, gave me a first and fascinating introduction to the fauna of Ngorongoro and the Serengeti. In South Africa I was again most fortunate to have Ian Player as mentor and guide on an extended tour of the Zululand game reserves. I am also especially grateful to Con and Molly Benson for much kind help during a stay at Chilanga in Zambia: and to John and Daphne Ball for hospitality and many kindnesses shown on my frequent visits to Kenya in recent years.

Other former travelling companions – in Mozambique, Uganda and Zambia – I am no longer able to thank personally: Leonard Bushby, Curator of Insects, and Jack Lester, Curator of Reptiles, Zoological Society of London; Roy Wyndham, Game Department of Uganda, who in tragic circumstances was killed by a lion; and Peter Maclaren, Fisheries Officer, who died after being attacked by a crocodile in the Zambesi.

The various research expeditions would not have been possible without financial grants, gratefully received on different occasions, from the University of Cambridge, the Royal Society, the Percy Sladen Trustees, the Government of Northern Rhodesia, East Africa Fisheries Research Organization, Uganda National Parks and the Zoological Society of New York.

I am indebted to Methuen and Co Ltd for kindly allowing reproduction here of the photograph on Plate 54, and for the figures on pages 158, 162–6, 167b, 177 and 182, which first appeared in *Adaptive Coloration in Animals*; to The Fountain Press, for photographs on Plates 6, 52, 53 and 55, and for the figures on pages 20 and 189, which are taken from *Zoological Photography in Practice*; and to Macmillan and Co Ltd for the figures on pages 19, 51, 70, 74, 82, 92, 120, 131, 169 and 206 from *Uganda in Black and White*. I am also grateful to the Editors of *The Royal Academy Illustrated* (1968) for the figure on page 207. My great indebtedness to the authors of books and papers consulted when preparing the present work is, I hope, adequately acknowledged by references in the text.

I

The Animal Community

Where order in variety we see
And where, though all things differ, all agree.
Alexander Pope

A TRAVELLER on his first visit to the great National Parks and wilderness areas of Africa will at once be impressed, and perhaps bewildered, by the extraordinary diversity of habitat, of vegetation and of animal life he will encounter.

If he were to take the launch trip to Kabalega Falls, he would find basking crocodiles and hippo schools dominating the shore and shallows; herds of elephant and buffalo will be seen coming to water; and an astonishing variety of birds – herons, egrets, storks, ibises, weavers, waders, Fish Eagle, Egyptian Goose and Pied Kingfisher.

In the evergreen forest, where great trees like *Busanga cecropioides* support a dense canopy more than a hundred feet above the ground, he will hear the raucous calls of Black and White-casqued Hornbills or the croaking notes of the Great Blue Turaco; or catch a glimpse of Black and White Colobus, their tails hanging like bell-ropes from the branches, or of Bush Baby, fruit bats, tree-frogs and brilliant swallowtail butter-flies.

In the Western Desert of Egypt he would find a totally different world where many members of the fauna – gerbils, jerboas, Fennec Fox, skinks and vipers – are fossorial; and where birds such as the Hoopoe Lark, sandgrouse and coursers are camouflaged and rarely seen until they have taken flight.

How different again is the vast Serengeti steppe – home of the famous Wildebeest herds, and of zebra and gazelle in their tens of thousands; of lion, Spotted Hyaena and jackals; and of Griffon Vulture, Secretary Bird, Kori Bustard, Ostrich and Capped Wheatear.

Confronted by all this rich diversity of life – each habitat supporting its characteristic community of plants and animals – a visitor can too easily overlook the concept so well expressed in the quotation which heads this chapter.

Tropical forest; the huge tree with buttress roots is *Musanga cecropioides*

Structure of the animal community

Every major habitat, together with the plants and animals it supports, constitutes a self-sustaining system which is organized in a particular way. The organization is basically the same in totally unlike habitats. Each ecosystem is essentially an association of producing, consuming and decomposing organisms.

The trees, grasses and other green plants are the primary producers. By the alchemy of photosynthesis they alone can transmute simple inorganic substances into complex proteins and other organic compounds which are the stuff of life. Directly or indirectly, all animals are dependent for their food upon plants. As Sir Arthur Shipley wrote in his delightful book entitled *Life* (1923): It is chlorophyll 'which makes the world go round'.

In the second category of organisms – the consumers – are the animals which according to their food-habits can be classified broadly as herbivores, predators and parasites, and scavengers. Secondary producers – herbivores at the bottom of the food hierarchy – are often very small and numerous, such as copepods, aphids, ants and so on. These are basic industry animals that provide food for the various grades of carnivores.

At the head of the food chains are terminal species. These are the master predators, or animals of formidable size or powers of defence which have nothing to fear from predatory attack. Such are the large Carnivora, the great ungulates, many storks and birds of prey, crocodiles, tigerfish and Nile perch.

Between the basic industry animals and the terminal species is a bewildering variety of interconnected forms – many of them insectivorous invertebrates and vertebrates such as the mantids, tree-frogs, or chameleons – which feed upon creatures smaller and more numerous than themselves, and which are in turn the prey of larger and less numerous hunters whose status is near or at the summit of the food hierarchy.

The third category, the decomposers, consists of organisms that break down dead organic material – excrement, exuviae, wood and corpses. This vital work of recycling the contained carbon, nitrogen and minerals is carried out by fungi and moulds, putrefying bacteria, and by such animals as termites, mites, earthworms, scavenging vertebrates, and by scarab beetles. An example of the latter is the giant dung beetle *Heliocopris dilloni* which feeds on elephant dung and provides for its offspring by laying its eggs in a buried cache of the same material.

Heliocopris dilloni, the giant dung beetle

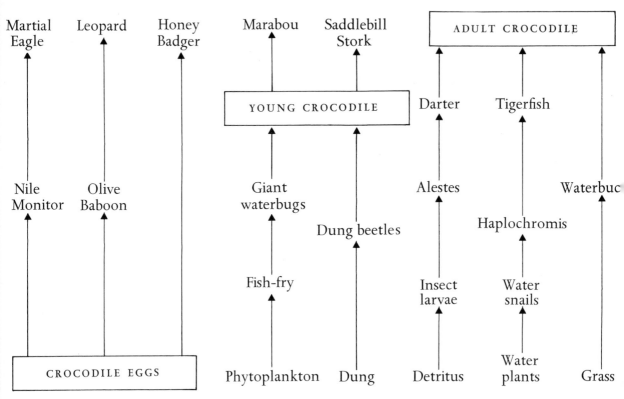

Martial Eagle   Leopard   Honey Badger   Marabou   Saddlebill Stork   ADULT CROCODILE

YOUNG CROCODILE   Darter   Tigerfish

Nile Monitor   Olive Baboon   Giant waterbugs   Alestes   Waterbuc

Dung beetles   Haplochromis

Fish-fry   Insect larvae   Water snails

CROCODILE EGGS   Phytoplankton   Dung   Detritus   Water plants   Grass

16

Nile Crocodile

The food web

A few strands selected from the extremely complex web of relationships in which the Nile Crocodile is involved either as prey or predator are shown opposite.

In the first months of life the hatchling crocodiles prey extensively upon giant waterbugs (Belostomatidae), nymphs of dragonflies (Gomphidae, Libellulidae), various water beetles (Dytiscidae, Hydrophilidae) and fresh-water crabs (*Potamonautes*). These invertebrates in turn feed upon young frogs and fish-fry, whose food consists variously of vegetable matter, zooplankton or detritus. The young crocodiles themselves fall prey to many enemies ranging from the soft-shelled turtle *Trionyx* and monitor lizards to such waterbirds as Marabou, Saddlebill Stork,

Encounter

Great White Egret and Fish Eagle. Meanwhile adult crocodiles include in their diet some of these eaters of their young; and after death the crocodile's body is attacked by different grades of scavengers and decomposers – vultures, blowfly maggots, river crabs and by cannibal crocodiles.

A glance at such relationships, which have been considered in more detail elsewhere (14), makes one realize how right was the Frenchman who said that life was summed up in the conjugation of the verb 'manger': *Je mange, Tu manges, Il mange, Nous mangeons, Vous mangez, Ils mangent:* and its fearful corollary – *Je suis mangé, Tu es mangé, Il est mangé, Nous sommes mangés, Vous êtes mangés, Ils sont mangés.* This is the process which, in more prosaic and precise language, the ecologist terms the 'conversion cycle'.

Johnston's Tree-frog, Mozambique

Niches and analogies

We have seen that animal communities in very different habitats resemble each other in their essential ground plan. This resemblance in the organization of ecosystems is more apparent when we consider the role played in the community by particular species. The concept of the *niche* was first admirably developed in 1927 by Charles Elton in his

Jackson's Chameleon, Kenya

classical work *Animal Ecology*. 'The "niche" of an animal means its place in the biotic environment, *its relations to food and enemies*' (22). There is often, as Elton has pointed out, an extraordinarily close parallelism between niches in widely separated communities.

There is a niche that is filled by 'cleaners' which rid reptiles of ecto-parasites. In Uganda this office is performed by the Spurwing Plover, which removes tsetse flies and leeches from the Nile Crocodile. In Aldabra a gecko *Phelsuma* feeds on mosquitos which swarm on the island's Giant Tortoise. In the Galapagos a scarlet Grapsid crab takes ticks from the marine lizard *Amblyrhynchus*.

Again, in tropical regions there is a niche for mammals which, as gastronomic specialists, feed almost exclusively on ants and termites. In Africa this habit is practised by the Aardvark (*Orycteropus*) and in Brazil by the unrelated Giant Ant-eater (*Myrmecophaga*). Both are fairly large terrestrial forms which rip open the nests with powerful claws and take the insects on their long extensile tongue. In the trees the same role is played by the African Long-tailed Pangolin (*Manis*) and by the South American Two-toed Ant-eater (*Cyclopes*): both animals (belonging to different orders) are well equipped for this strange mode of life, being strong-clawed and toothless, and having a tubular mouth, long sticky tongue, and prehensile tail.

The sunbirds in Africa and hummingbirds in the Americas provide another example of close resemblance between animals that occupy a similar niche in different parts of the world. In both groups the beak and tongue are modified to probe for and to extract nectar and tiny insects from flowers. It happens that the birds also share other characters, being diminutive in size and having brilliant iridescent plumage, though they belong to different avian orders: humming birds are more nearly related to swifts than to sunbirds.

Another niche is that occupied by eaters of birds' eggs. Such predators are found in every habitat and in every geographical region: Greater Black-backed Gull, Carrion Crow, Magpie, Jay and Marsh Harrier in Europe, or Great Skua and Sheathbill in the Antarctic are examples. For animals of high latitudes egg-eating is of course a seasonal occupation. But in the tropics, where birds of different species are found nesting all the year round, it is possible for an animal to become entirely addicted to eggs. One such is the very remarkable African egg-eating snake, *Dasypeltis scaber*. Only three feet in length and proportionately slender in girth this snake is even able to swallow a hen's egg. Peculiar modifications of the 'neck' vertebrae, some of which have ventral spines that penetrate the dorsal wall of the gullet serve, like a tin-opener, to slit the shell as it passes towards the stomach, enabling the reptile to swallow the contents without loss, and to regurgitate the collapsed shell.

Termites as basic industry animals

Anyone who has flown from Entebbe to Kabalega Falls will have noticed that over vast areas the ground is pock-marked with a kind of pink

Termite castle, Buligi, Uganda

rash. In a vertical air photograph the ground pattern shows a stipple of white spots strongly differentiated from the dark texture of grass. The spots tend to be fairly regular in distribution, at intervals of a hundred yards or so. Each spot is a termite mound. Owing to trampling of animals, many of which like elephant and buffalo use the mounds as rubbing posts, the adjacent ground is bare and this makes them all the more conspicuous. For mile after mile the insect fortresses dominate the scene and the fact is borne in upon the observer that these insects are indeed a key form of animal life in wilderness areas.

In their habits termites, or 'white ants', show closer similarity to ants than to other insects, since both live in highly organized societies in which individuals are differentiated into castes. But the complex social organization has evolved independently in the two groups. Ants are members of the stinging or aculeate Hymenoptera – a group which includes the wasps, bees and bumble bees: they have a thick cuticle; a slender 'waist' between thorax and abdomen; and development is of a kind known as 'indirect' – by the stages egg, larva, pupa and adult. Termites belong to the order Isoptera, a much more primitive group with affinities to the cockroaches. They have a thin cuticle, are stout-bodied, lack a sting, and develop directly from the egg through stages which resemble the adult.

22

It seems remarkable that these apparently frail and vulnerable insects have so successfully spread their teeming empire throughout the warmer parts of the world. In his fascinating book on the biology of termites, Howse (34) writes: 'At first it is hard to imagine a creature more poorly adapted to life on earth than the typical termite. Blind, slow, weak, with a soft skin, unable to stand sunlight or changes in temperature or humidity, it nevertheless survives floods, droughts, ants and other prowling insect predators by enclosing itself in a hard-shelled fortress. It builds elaborate tunnels to its sources of food, in order to reduce to a minimum contact with the harsh conditions of the outside world.'

The mounds differ much in architecture according to circumstances and from species to species. Sometimes they become gigantic so as to alter the character of the savanna. The newly-constructed mounds of *Macrotermes* do not support vegetation. During the rainy season, soil is eroded from the mound externally, while the termites maintain it from within. Meanwhile the weathered soil accumulates at the base of the mound, giving rise to a hillock. Over the years new termite colonies take possession of the hillock and a succession of mounds arise on the same site. Further accumulations of eroded material raise the height and girth of the hillock until after perhaps centuries it becomes a hill. In Zambia old nests of *M. bellicosus* attain a height of twenty-five feet and a basal diameter of over a hundred feet, and are covered with trees and scrub which testify to their great age.

Unlike bee and wasp societies which are matriarchal, with sterile females forming worker castes and drones which are short-lived, termite societies are centred round a royal pair, the king and queen. The castes are juvenile forms and include workers, soldiers, and potential reproductive individuals of both sexes. Specialization and division of labour are the order of the day.

When danger threatens, whether from damage to the mound or from attacking ants, the workers retire into the fortress and the soldiers swarm to its defence. The soldiers are aggressive and are variously armed in the different species. An entomologist named Escherich has described how at his first meeting with *Macrotermes bellicosus* in Ethiopia he received a dozen deep cuts in his fingers from which blood came in streams. These wounds were of course inflicted by the mandibles. In some species the mandibles meet like forceps; in others they overlap, to cut like shears; others again spring open with stored energy like the snapping of fingers, and with enough force to decapitate a raiding insect. In *Nasutitermes* chemical warfare is employed by highly specialized soldiers –

the 'nasutes'. These forms lack mandibles, and have the head-capsule produced forwards into a nozzle: from this can be shot a secretion which becomes viscous in the air, to envelop an adversary with a sticky web.

Workers perform the carrying, building, cementing, foraging, gardening and other maintenance duties; and attend the royal couple. The queen in particular requires service by scores of workers: as noted by one observer, 'rows of them are arranged all round her body caressing her with their palps; and there is a constant stream of workers to and from her anal region.' Thus pampered, and with her bloated white body vastly distended, the queen becomes virtually an egg-laying machine. The queen of *Macrotermes natalensis* is said to produce 36,000 eggs a day, or some 13,000,000 in a year. By some African tribes the queen termite is believed to possess medico-magical properties, and is eaten by the women.

Termites are of course notorious as destroyers of wood, though the different species feed on a wide range of substances containing cellulose – humus, manure, cotton, paper and cardboard. Harvester termites even forage above ground for grasses, which are cut up and transported to the nest. Being themselves unable to digest cellulose, termites harbour or cultivate symbiotic organisms which perform this function for them. Some rely upon the presence of protozoa in the intestine; others upon an intestinal microflora of bacteria. Others again have a symbiotic association with fungi which they cultivate in the nest. Having extracted the digestible constituents of their food, the workers pass out in the faeces the indigestible remains. This material they mould into elaborate 'combs' – the so-called fungus gardens, where the fungi in turn break the cellulose fragments into simpler substances.

Termites as food

Termites have an important part to play in the ecology of any area where they occur. In the first place, they form a main, subsidiary or occasional prey for a wide variety of animals – frogs, lizards, mammals and birds of many kinds. Notable among termite-eaters are the specialists – the Aardvark (*Orycteropus afer*), and the pangolins of which there are several African species; and their counterparts, the ant-eaters of South America. The myrmecophagous habit demands various structural modifications of the body. Such animals do not require teeth: the Aardvark has reduced dentition; the others are edentate. All have powerful claws, with which they break open the nests. All have the skull greatly elongated, with a narrow mouth, and very long tongue which is extensile, coated with a sticky secretion from the well-developed salivary glands, and admirably adapted to probe the termite galleries. A termite mound with a gaping hole in the side is a common sight: this is the work of an aardvark. The animal itself is rarely seen, being strictly nocturnal. Another mammalian predator is the Aardwolf (*Proteles cristatus*), a relative of the hyaenas. According to recent studies by Kruuk and Sands (39) in the Serengeti, this animal is highly selective and concentrates upon a single species – *Trinervitermes bettonianus*. It does not use its paws for digging but walks slowly over the ground, locating foraging insects at the surface by scent or hearing.

When after rain the winged termites are emerging, predation is very high: swifts, swallows, bee-eaters, bulbuls, starlings, flycatchers, hoopoes, barbets, shrikes and birds of prey are among those attracted to the scene. If the flight is dense and continues as a locally abundant food supply, birds are drawn to the feast in increasing numbers. When camped at Namsika below Kabalega Falls I have seen about fifty Black Kites wheeling and swooping on the fluttering insects, which of course they take one at a time with the foot, to be transferred in flight to the beak. On another occasion, when after a rain-storm termites were emerging from holes in the ground, a considerable flock of Buff-backed Herons arrived, to run about in the way hens do at feeding time. Even Marabous find it worthwhile to gather such small morsels.

The nuptial flights usually occur during the rains. The Pennant-winged Nightjar (*Semeiophorus vexillarius*) which feeds largely upon winged termites, enjoys a double advantage, being a trans-equatorial migrant. It breeds during the early rains in November in the southern savanna, and then flies north in February to spend its off-season in southern Sudan and Uganda where again the rains make available its favourite food.

Another opportunist is the so-called 'rain-frog' *Breviceps*. These strange animals, rotund in body with short legs and the head profile of a pekinese, are not seen in dry months when they hide away aestivating – two dozen or more huddled together in a hole which provides a moist microhabitat. But after rain they appear from their hiding place in great numbers, to congregate round the exit holes of emerging termites. At Ndumu in Zululand after rain in late October I have seen numbers of these frogs, together with a large ranid *Rana adspersa*, absolutely cramming themselves with termites that were emerging from holes in sandy ground at the roadside. In his delightful book *Zulu Journal* Professor Raymond Cowles (16) refers to brevicipitid toads gathering to the feast wherever lights attract the insects: 'They gorge until their stomachs and finally their mouths are full, and one can see them reaching up with

Termite exit vent

stubby forelegs to tuck in still more termites and to rub away the
protruding wings that seem to tickle their lips.'

Termites are palatable to man, and are eaten alive, or fried in their
own fat and made into a kind of cake. On the Victoria Nile a few miles
below the Owen Falls I once came across a party of natives who were
collecting termites. They had removed the superstructure of a mound,
and most of the galleries leading down into the nest they had blocked
with clay. To some half dozen remaining exit holes they had fitted
chimney-like cylinders made from the material of the nest which they
worked with water into clay, adding a few struts for support. Four boys
had planks of wood arranged round the base of the termitarium, and
these they were beating rhythmically with short stout sticks with such
vigour as to make the ground vibrate. The purpose of this was to simulate a torrential downpour and so to precipitate a nuptial flight. As we
watched, the insects began to emerge from each chimney; those escaping the men were swept up by swallows before they had fluttered ten

feet into the air. Some species swarm at night; and fires are lit to attract the flying insects which can then be caught in handfuls. We may note in passing that Smeathman, who in 1781 published the first scientific account of African Isoptera, considered roasted termites superior in taste to shrimps.

Use of the mounds by other animals

Animals of many kinds use termite mounds variously for shelter. The gaping holes excavated by aardvarks are utilized by warthogs as a nocturnal retreat: before entering, the animal turns round and goes in backwards so that if danger threatens it can defend itself or bolt. Old mounds become honeycombed with deep underground passages spacious enough to be used as dens by hyaenas; and it is in such places that one can see – as for example near Keekerok in the Mara plains – several females tending their litters of different ages. Packs of Banded Mongoose, as recorded by Ernest Neal (59), similarly utilize termite mounds as headquarters.

The African Rock Python (*Python sebae*) has been seen in Uganda to take up residence in a mound, using it as a shelter when the female is coiled round her eggs. In Zululand Cowles made the interesting discovery that the Nile Monitor habitually uses a mound of *Nasutitermes trinerviformis* as an incubator for its eggs. The clay-and-saliva-cemented home, baked by the sun, is very hard, but conditions are right for egg-laying when the rains have softened the outer layer of the nest. The monitor scrapes a hole and lays her eggs deep in the warm interior of the mound; these are soon covered over with clay by termites repairing damage to the superstructure. About nine months later the lizards hatch and work a vertical passage through which they escape to freedom.

The role of predators

Predators play a key role in the life of the community; and, paradoxically, they tend to benefit the species on which they prey. In the first place, by a selective process they promote the survival of the fit. Birds of prey and carnivorous mammals are quick to detect and pull down sick, weak or old individuals immediately they lag behind their companions. In his original and informative *Pirates and Predators*, Colonel Meinertzhagen reports that hawks, and even crows and ravens, are very quick to spot a 'pricked' bird: 'Any bird accustomed to the protection of the flock,' he writes, 'falls an easy victim to predators without that protection... There is little doubt that predators exercise a selective effect in choosing abnormal birds whether in colour, health or some physical or mental weakness.'

From the field studies of Kühme (41) and the van Lawick-Goodalls (44) it seems clear that the hunting technique of the African Wild Dog enables members of the pack to select from the herd an antelope that is weaker or slower than his fellows. Hard proof of this type of selection is not easily obtained; and a study of the feeding habits of cormorants, carried out by van Dobben in Holland, is of special interest in the statistical evidence it provides. The birds' main prey in the study-area were roach. Examination of stomach contents showed that 30 per cent of the fish captured by the cormorants were infected with a tape worm: whereas the incidence of worm infection in the free roach population was only 6·5 per cent.

Secondly, predators play a vital part in regulating populations and so preventing ungulates from increasing beyond the carrying capacity of their habitat. A classical example of the imbalance which can result from the elimination of wild carnivores is afforded by the history of the deer population on the Kaibab plateau of Arizona. Originally pumas and wolves seem to have maintained the population well below the capacity of the range at about 30,000 head. Between 1907 and 1939 over 800 puma and 7,000 coyotes were killed, and wolves were exterminated. After the shooting the deer population rapidly rose, to outgrow the winter food supply; and by 1924 had reached 100,000 animals. Then followed a catastrophic drop within one year to 40,000, and by 1939 the population stood at about 10,000 deer living on a depleted range.

In Africa awareness of the problem began in the early thirties when Stevenson-Hamilton, for many years Warden of Kruger Park, reported an increase of Impala, which he related to the almost total disappearance by 1933 of the park's formerly large packs of Wild Dog. Today in Uganda it is probable that the excessive number of baboons which forage in large troops on the Acholi and Bunyoro banks of the Victoria Nile may be attributed to growing scarcity of their main enemy, the Leopard.

Thirdly, predators – and especially the pack-hunting Wild Dog and Spotted Hyaena – tend to enforce beneficial movement among the herds of ungulates, driving them off a grazed area to new grounds, and generally fulfilling the same function as wolves in northern latitudes. In winter the American White-tailed Deer tend to hang about, or 'yard', in one area; wolves prevent yarding, split up the herds, and so help to conserve the habitat.

Reactions of the prey

Many writers have commented upon the strange lack of concern which herbivores show in the presence of enemies when the latter are not actually hunting. Prey animals give one the impression that they are aware of the predator's intentions, and react accordingly. 'It is not an unusual sight', writes Stevenson-Hamilton, 'to see in open country an old bull wildebeest grazing within a hundred and fifty yards of some resting lions.' The same authority recounts how a ranger came on a lion and two lionesses taking their midday siesta less than a hundred yards from a troop of zebra. 'Neither party was taking the least notice of the other . . . One lion was lying on its back with all four legs in the air, like a cat before the fire, while the zebras were standing about apparently half asleep. The lions must have been clearly visible to them. Incidents such as this illustrate the perfect understanding which the wild creatures have of each other's ways, and how far man is from a clear comprehension of them' (74).

Warning signals which denote alarm are infectious, and are recognized not only by companions but by members of other and often quite unrelated species. Crocodiles, by no means as sluggish and dull-witted as they seem, are extremely quick to take the cue not only from their regular avian companions the Spurwing Plover and Water Dikkop, but also from the Egyptian Goose and other birds that frequent the basking grounds. The alarm honking of a goose or the bark of a baboon is enough to send the reptiles stampeding to the water.

There is no question that birds warn game of the approach of man. When oxpeckers fly up, uttering their harsh, scolding notes, the animal they have left is at once alerted, as a man would be by an alarm clock. Meinertzhagen thus records the reactions of a Grevy's Zebra which, caught as a foal, had been over a year in captivity and was very tame and gentle. 'He had seven yellow-billed birds on him; as I approached they rose and yelled; this threw the Zebra, who normally would come for sugar, into a panic and he raced round as though the devil were after him . . .'

Among vertebrate animals generally, each species has its own language of signals, which is unintelligible to or ignored by members of another species. I have referred to the specifically distinct nature of animal communication in Chapter 9. But, as Estes (23) has pointed out, alarm signals are the exception to the rule – they are recognized as a warning not only by herd-members but by associated species. 'Any ungulate that stands erect or stares in one direction, or suddenly breaks into flight, alerts all nearby game.' The alarm snorts, often accompanied by stamping with a foreleg, are 'basically alike and mutually alarming'. Thomson's Gazelle also signals visually by twitching the flank skin, or by 'stotting' just before taking to flight, and so semaphores a warning to all animals in the vicinity.

To witness a kill is for the human observer a distressing experience, especially when – as with prey of the Wild Dog – death comes slowly. It does seem likely, however, that in certain circumstances the terror and physical pain of the wounded animal are to some extent mitigated by traumatic shock.

Henry Fosbrooke speaks of Grant's Gazelle sitting in a state of equanimity whilst wild dogs proceeded to disembowel them, as though they were under an anaesthetic. In Ngorongoro a terrier belonging to the same writer was taken by a leopard and severely bitten in the small of the back so that the liver and kidney were protruding. 'It was obvious', he writes, 'that the dog felt no pain. He was in a state of deep shock, and pathetically wagged his tail and licked our hands as we patted him' (26).

Although the human spectator naturally feels pity for the animal that is dragged down in the hunt, we do well to remember that, as George B. Schaller remarks in his *Serengeti – a Kingdom of Predators*, 'Predators are the best wildlife managers'. Owing to the mistaken belief, formerly held, that they harm the species on which they subsist, the predators themselves have suffered most from persecution by man; and today in many parts of Africa it is the hunting rather than the hunted species whose continued survival is threatened.

Ants and acacias

Everyone who has visited Nairobi National Park or driven through the thorn bush areas of the Serengeti will have noticed the abundant small acacia bushes that are armed with long white spines and decorated all over with blackish objects about the size of a golf ball. On closer inspection the latter turn out to be hollow galls, each inhabited by a colony of very aggressive ants which emerge through holes when the bush is disturbed – to swarm over the observer and sink their mandibles into his person.

The bush is *Acacia drepanolobium*, the Whistling Thorn – so named from the flute-like sound of the wind in the galls – and the ant, a species of *Crematogaster*. It is not known whether development of the galls is induced by the ants or whether the bush would produce them in any case. But

Whistling Thorn, showing protective armament of spines, and domiciles of the commensal ant *Crematogaster*

the association is a regular one between insect and plant, and is an example of a widespread phenomenon.

These relationships have been discussed in a paper entitled *Ants, Acacias and Browsing Mammals* by W. L. Brown (6) and more recently by Foster and Dagg (28). Many plants carry extra-floral nectaries which contain a sugary solution attractive to ants, and also bear special shelters, or domiciles, in which the ants live. It seems probable that the relationship is a mutually advantageous one: the plant encourages the presence of ants by supplying food and home; and the ants afford some protection, just as spines do, from excessive damage by browsing ungulates.

Despite the presence of ants, and spines, the giraffes browse on Whistling Thorn and other acacias. Deterrence is never absolutely, but only relatively, effective. Observations have shown that when a giraffe has browsed for about one and a half minutes, by which time the ants will have been stirred up to great activity, it moves on to the next plant. In this way the damage from browsing tends to be spread through the tree population. Aberrant whistling thorn bushes which lack galls are always heavily browsed.

It is interesting to note that whereas in Africa, the home of so many browsing ungulates, species of *Acacia* are mostly spine-bearing and myrmecophytic, trees of the same genus in Australia – which lacks an effective fauna of browsing mammals – are without spines, ant domiciles and extra-floral nectaries.

Protective nesting associations

An animal's life is not dominated – in the struggle for security and subsistence – by interspecific warfare. Very often there have evolved other close relationships between unrelated species, not of enmity but of assistance. In some the benefit is one-sided; but in many cases both partners benefit from the association, which in effect is one of mutual assistance, or even of interdependence.

I have referred in Chapter 6 to the habit of Cattle Egrets, which persistently attend Elephant, Buffalo or other game as 'beaters' to feed on insects flushed from the grass. At Buligi in Uganda one can often see these birds, in ones or twos, walking close beside Uganda Kob, and feed-in the same way as starlings do when going with sheep in an English meadow. In Uganda the Piapiac, a relative of the starling, is also a regular attendant of Elephant and Buffalo.

Some birds acquire adventitious protection from enemies by associating with formidable stinging insects – habitually building their nests close to colonies of aggressive ants, bees, wasps or hornets. These nesting

Cape Buffalo and Piapiac

partnerships were first studied in South America and the West Indies by J. G. Myers (58) who showed that the association between orioles (Icteridae) and insects was not a casual one, but regular; and that the nesting partners chosen by the birds were typically the most vicious members of the group – for example, the wasps *Polybia* and *Polistes*. Sometimes the purse-shaped birds' nests were built so close that the homes of bird and insect rattled together when blown in the wind. Yet the wasps, quick enough to attack a mammalian intruder, did not molest their avian neighbours.

Similar associations are known from the tropics of Australia and Africa, and have been recorded of species belonging to several families. In 1950 P. I. R. Maclaren (49) found that in Nigeria the Bronze Mannikin (*Spermestes cucullatus*) regularly built in close proximity to one or more nests of a vicious red ant (*Oecophylla smaragdina*), and the bird's definite preference for nesting in trees already occupied by ants' nests was statistically demonstrated. In East Africa the same mannikin has been seen to associate with the wasp *Ropalidia nobilis*. The Blue Waxbill (*Uraeginthus angolensis*) is another regular associate with wasps when nesting. In Cameroon the Red-headed Lovebird (*Agapornis pullaria*) breeds in a hole scooped out of an arboreal termite's mud-nest. One such nest was occupied by large ants that savagely attacked a boy who climbed the tree. In these associations with stinging Hymenoptera, the advantage is all on the bird's side: it is the bird that seeks and benefits from the partnership. Strangely enough, there exists complete mutual toleration between the partners: the young birds are not attacked by the insects, and the nesting birds leave the insects alone.

Many cases are known of small, inoffensive birds habitually nesting near the nest of an aggressive raptorial species. The Black-headed Weaver (*Ploceus cucullatus*) – sometimes called the 'village weaver' – habitually nests with birds of prey, or with man. In some weavers the 'protector' is a Fish Eagle or Marabou.

I came across an example during a visit in 1956 to Mkuzi Game Reserve in Zululand. The nest of a Wahlberg's Eagle (*Aquila wahlbergi*) built in the upper branches of a knob-thorn tree (*Acacia nigrescens*) was closely surrounded by a colony of nests of the Spotted-backed Weaver (*Hyphantornis spilonotus*). Most of the weavers were still building when the eagle's nest was complete and the eagle was already incubating its single egg. This was in a well-wooded area in which innumerable other trees were available: yet the weavers – the only *H. spilonotus* colony in the vicinity – had chosen the one tree that was occupied by a large bird of prey.

Bird-mammal feeding associations

Oxpeckers, or 'tick-birds', of which there are two species, are related to starlings (Sturnidae) and are only found in Africa. Both Yellow-billed (*Buphagus africanus*) and Red-billed Oxpecker (*B. erythrorhynchus*) patronise a wide range of ungulates. All the larger species, except the Elephant, receive their attention, and parties are commonly seen on Black Rhinoceros, Eland, Giraffe, Buffalo, Zebra, and on Warthog. The relationship with Rhinoceros is most intimate.

The host animals generally show surprising tolerance of the birds though – as one may often witness at Treetops – bushbuck resent their presence and do their best to shake them off. 'But,' as Moreau remarks, 'they are very difficult to get rid of, and if forced to leave a beast they immediately settle on another. In fact their movements resemble those of blow-flies on meat in their pertinacity . . .' Meinertzhagen saw warthogs assisting oxpeckers in their search for parasites, 'turning the head slowly from side to side so that they could get at the creases in the neck.'

In accordance with their highly specialized mode of life, oxpeckers show a number of structural adaptations. The short curved claws are needle-sharp and enable the bird to cling to the hide of a galloping animal, and to run all over the host's body in any direction – like a woodpecker on a tree-trunk – up and down the leg or flank, beneath the belly, or on the head to inspect ears and nostrils. The tail feathers of *Buphagus* are stiff in texture, like those of a woodpecker, and are used as a support. The beak is laterally flattened and can be used in a scissoring movement when held flat against the hide of the host animal.

Blood-gorged ticks form the main diet; but the birds also take insects such as the blood-sucking fly *Stomoxys*, and they attack open sores to feed on flies, scar tissue and blood. In 1969 C. A. Spinage (73) was able, by using a special technique, to make quantitive assessments of ectoparasites present on large skins: the numbers found greatly exceeded expectation. For example, one side of a ten year-old waterbuck's body, with the legs, carried 2,033 ticks and 247 lice. From such observations it seems clear that oxpeckers are in fact exploiting a prolific food source, and one which is hardly, if at all, tapped by other birds. An investigation in Tanzania showed that of 58 oxpeckers examined, 55 contained over 2,000 ticks in their stomachs. The association is specialized and, for the birds, obligatory. They obtain all their food from the animals they haunt; they use their host's body as a perch and refuge, and as a place for display and copulation; and its hair as material for lining their nests.

The arrangement, however, is not one-sided. For the birds act as senti-nels, warning game with harsh, scolding notes as soon as a human being comes near.

The association of the birds with Black Rhinoceros is particularly close; and the host animals immediately respond to the birds' alarm calls. There are records of oxpeckers staying with their host hours or even days after it had been shot. R. Gordon Cumming, who hunted in South Africa in the 1840s, was often disappointed in his stalk by the ever-watchful birds. He relates that when he had shot rhinos at night, the oxpeckers would remain with them until morning, and on his approach would then exert themselves to awaken their companion from his deep sleep before themselves taking flight.

Commensal birds and the Nile Crocodile

Almost parallel to the commensal association of oxpeckers with rhino-ceros is the close relationship between certain birds and the Nile Croco-dile. Classical accounts of the bird which was said to attend the crocodile in Egypt are well known. According to Herodotus: 'The crocodile is in the habit of lying with its mouth open, facing the western breeze: at such time the *Trochilos* goes into his mouth and devours the leeches. This benefits the crocodile, who is pleased, and takes care not to hurt the *Trochilos*.' The identity of the *Trochilos* was much debated in the last century. Among modern ornithologists, and despite the observations of reliable witnesses, there has been a tendency to dismiss the story as a myth.

In Uganda the Spurwing Plover (*Hoplopterus spinosus*) is the reptile's constant companion on all the favoured basking grounds, and may be seen running on, flitting over or standing close beside the sprawled bodies (14). Crocodiles immediately respond to their shrill alarm call – 'quick-quick-quick', which is sufficient to alert the whole congregation and to start a stampede to the water. Another commensal is the Com-mon Sandpiper. These waders run up to a crocodile as soon as it has hauled-out on to the bank, and will systematically work round the body in quest of ectoparasites. In Zululand these birds have been seen to pick up food from a crocodile's mouth and pull a leech from the mucosa, and to tug leeches from the gular shields. A third regular associate is the Water Dikkop (*Burhinus vermiculatus*); in the breeding season, as observed in Uganda and Zululand, one or more pairs are frequently found on the rookeries. Like the plover, they play the role of watchdog and give the reptiles timely warning of danger.

On the lower reaches of the Nile the crocodile's companion was the Egyptian Plover (*Pluvianus aegyptius*). The German naturalist A. E. Brehm, who travelled in Egypt and the Sudan in the 1850s, says of this plover: 'Without the slightest hesitation it runs around on a crocodile as if it were just a bit of green lawn, pecks at the leeches that are bleeding the reptile, and even has the courage to take parasites adhering to the gums of its gigantic friend. I have seen this on several occasions' (4). Confirmation of this account is given by Colonel Meinertzhagen (53), who in the Sudan watched the plover perch on the jaw of a crocodile, inspecting and pecking at something in the mouth. On the Kafue River in Zambia, the same role is played by the Blacksmith Plover (*H. armatus*).

Honey guiding

Finally, we must mention here the strangest of all commensal relationships, which has its origin in the specialized feeding habits of certain African honey-guides. The Black-throated Honey-guide, most appropriately named *Indicator indicator*, is a small woodland bird, starling-sized and subfusc in coloration, which feeds on insects, but has a predilection for the larvae of wild bees and especially for wax of the honeycomb. The bird is quick to find a bees' nest but, being unable to reach the booty unaided, has developed the unique behaviour of seeking help from the Honey Badger, or from a man.

The bird begins by approaching the honey-gatherer and attracting his attention by its persistent display and harsh repeated churring notes. It then leads the way to the nest, flying in the given direction but always keeping only a few yards ahead of its companion. On reaching the hive the bird perches nearby and waits. After the honey-gatherer has raided the nest and departed, the bird enters and feeds on the wax of the broken comb. Its unusually tough skin appears to be impervious to bee stings.

Accounts of the instinctive guiding behaviour of *Indicator* read almost like a fairy-tale, and the strange association of bird and man seems to be so 'improbable' that it might easily be dismissed as legendary, were it not for confirmation by reliable observers. In 1955 Herbert Friedmann, formerly head of the Bird Department of the United States National Museum in Washington and President of the American Ornithologists' Union, made a detailed field study of the honey-guides, in the course of which he was himself guided by the bird on twenty-three occasions to a place where honey was to be found.

This wonderful commensal association is tragically disappearing from Africa before the advancing tide of development. Once known to exist from Nairobi to Cape Province, it is now almost confined to the wilder parts of Zaire, Ruanda-Burundi and Kenya, where honey-hunters still follow their ancient practice. But the day cannot be far off when the honey-guide will display to sophisticated humans in vain; and the curious partnership that once existed between a free-living bird, and man, will again come to be regarded as a pleasant fable.

2
Herbivorous Animals:
The Ungulates

In not the earth
With various living creatures, and the air.
Replenish'd; and all these at thy command,
To come and play before thee? Know'st thou not
Their language and their ways? They also know,
And reason not contemptibly. With these
Find pastime.

Milton

THE term 'Ungulates', which embraces the hoofed animals, is a somewhat misleading one, in that the order Ungulata includes two groups
of animals not closely related, and which are now placed in two separate
orders – Perissodactyla and Artiodactyla. These are distinguished by the
structure of the foot. In the Perissodactyla or odd-toed ungulates – the
rhinos, tapirs, horses, asses and zebras – the central axis of the foot runs
through the third toe, which is the largest and is symmetrically flanked
by the second and fourth toes (when these are present). In the Artiodactyla or even-toed ungulates – a huge assemblage including hippos,
pigs, giraffes, camels, antelopes, oxen, sheep, goats and deer – the
central axis of the foot passes between the third and fourth toes, which
are symmetrical with each other: the second and fifth toes (when
present) are placed outside and behind the main toes, and generally do
not touch the ground. The condition in the two groups is known as
'mesaxonic' and 'paraxonic' respectively.

Convergent adaptations

The general resemblance in shape between such animals as rhinoceroses and hippopotamuses, or horses and cattle, are examples of
evolutionary convergence – a superficial similarity independently
acquired by animals which have adopted, and are adapted to, a similar
mode of life.

38

We may note in passing that convergence is a phenomenon widespread in nature; and it is found at different levels of affinity in the natural classification. For example, the Mole (*Talpa*), the Rodent Mole (*Spalax*), and the Marsupial Mole (*Notoryctes*) are remarkably similar in form and in numerous adaptations which fit them for fossorial life: yet they belong to distinct mammalian orders – Insectivora, Rodentia and Marsupialia. Convergent similarity may extend across different classes of a phylum, as in a shark, ichthyosaur and dolphin; in this trio the reptile and mammal are profoundly modified for pelagic life and both look like 'fish' though derived from terrestrial forebears. Even more curious are similar adaptations found in a humming bird and a Humming-bird Hawk Moth: here an insect and a vertebrate exhibit likenesses imposed upon them as flower-visiting hoverers.

The specialized members of the two ungulate orders show convergent similarities resulting from adaptation to cursorial life on hard ground. The metacarpal and metatarsal bones are lengthened, so that what seem to be 'knees' are really wrist and ankle joints raised high off the ground. By this lengthening of the leg's distal segment the stride is increased, while the powerful leg muscles are concentrated high up, thus quickening the pace by raising the bob of the limb-pendulum. All the limb joints (except the shoulder and hip) are hinged, so that movement is restricted to the fore-and-aft plane. The number of digits is reduced, and these are encased in solid hoofs·which alone touch the ground when the animal is running. The convergent similarity is brought even closer in the Giraffe and some other Artiodactyla by the total fusion of the metacarpals and metatarsals of the third and fourth digits to form a single bone in each leg which is the counterpart of the single 'cannon bone' in the horse's leg.

Ungulate variety

Every region has its diverse ungulate representatives. The distribution of some species, like the Klipspringer, is patchy, owing to the animal's specialised habitat requirements. Others like the African Buffalo are found over a wide range of terrain, from dense forest to open country. From the unique array of antelopes to be seen, and wondered at, in the game areas, it is only possible here to select a few for brief mention.

In the Buligi grassland of Uganda one may see the graceful Oribi, usually in pairs or family groups. Standing less than two feet at the shoulder, these little antelopes are distinguished by the round patch of

Topi bull, Ishasha, Uganda

black skin below the ear. They are grazers; and when flushed they gallop, dodge and leap through the long grass, uttering whistling alarm snorts and displaying the black-tipped tail as they go. In the same area Jackson's Hartebeest is common. A member of the herd may often be seen standing guard high and motionless on a termite-mound, its grave-looking face and excessively elongated head reminiscent of an El Greco portrait.

The closely related Topi replaces the hartebeests in south-west Uganda. More handsome than the 'kongoni', it is purplish-red in colour with blue-black patches on the face, flanks and legs. Topi are very gregarious, and the sight of large herds numbering several hundred animals grazing in the park-like savanna at Ishasha is a rewarding experience.

Ishasha is also noted for its Giant Forest Hogs. Nocturnal denizens of gallery forest and dense thickets near the river, they are less easily observed than the diurnal Warthog, which prefers open country. The massive muzzle, small pointed ears, coarse jet-black coat, and above all the impressive size serve to distinguish this elusive hog. Glimpsed at dusk as they move among the trees they might momentarily be mistaken for a party of Pygmy Hippopotamus.

A typical grassland antelope of Uganda is the Kob, whose head has appropriately been chosen as the crest of the Uganda National Parks. In 1957–59 Helmut Buechner, an American ecologist, carried out extensive field studies on reproduction in the Kob, and discovered that the species has a highly developed year-round pattern of territorial behaviour. Permanent territorial breeding grounds, roughly circular and up to two hundred yards in diameter, are located near water on a knoll or raised area of short grass, commanding good visibility. Within each main territorial area – which can be recognized by its closely-cropped grass and heavily-trampled ground – from twelve to fifteen males defend individual territories from about twenty to sixty yards across. Females move freely over the area and enter the territories for mating (7).

In the arid country of Karamoja, and the Northern Province of Kenya, are to be found some of the grandest antelopes – Beisa Oryx, Eland, Roan and Lesser Kudu: also some of the very attractive smaller species such as Klipspringer and Kirk's Dik-dik. The Kudu wears a cryptic dress and tends to escape notice, even when it is standing in full view, so effective is the combination of countershading and vertical white disruptive stripes on the body.

Klipspringers are found only on kopjes and rocky mountain slopes where their ability to leap and climb, sure-footed as a chamois, on precipitous places and precarious footholds is astonishing. Their non-skid hoofs are said to have the consistency of hard rubber. Dik-diks – large-eyed, spindle-legged, and little over a foot in height – are timid, secretive little animals occurring only in dry bush country: they particularly favour patches of thick undergrowth where aloes grow. A curious feature is the elongated nose which forms a proboscis used to strip leaves from low-growing bushes.

An impression of the rich diversity of Ungulate species found in the Serengeti National Park will be gained by reference to the accompanying list.

PERISSODACTYLA

EQUIDAE
Burchell's Zebra *Equus burchelli*

RHINOCEROTIDAE
Black rhinoceros *Diceros bicornis*

ARTIODACTYLA

HIPPOPOTAMIDAE
Hippopotamus *Hippopotamus amphibius*

SUIDAE
Warthog *Phacochoerus aethiopicus*
Bush Pig *Potamochoerus porcus*

GIRAFFIDAE
Masai Giraffe *Giraffa camelopardalis*

BOVIDAE

Eland *Taurotragus oryx*	Impala *Aepyceros melampus*
Lesser Kudu *Tragelaphus imberbis*	Grant's Gazelle *Gazella granti*
Bushbuck *T. scriptus*	Thomson's Gazelle *G. thomsoni*
Fringe-eared Oryx *Oryx beisa*	Red Duiker *Cephalophus harveyi*
Roan Antelope *Hippotragus equinus*	Blue Duiker *C. monticola*
Common Waterbuck *Kobus ellipsiprymnus*	Bush Duiker *Sylvicapra grimmia*
Defassa Waterbuck *K. defassa*	Suni *Nesotragus moschatus*
Bohor Reedbuck *Redunca redunca*	Oribi *Ourebia ourebi*
Mountain Reedbuck *R. fulvorufula*	Klipspringer *Oreotragus oreotragus*
Coke's Hartebeest *Alcelaphus buselaphus*	Steinbok *Raphicerus campestris*
Topi *Damaliscus korrigum*	Kirk's Dik-dik *Rhynchotragus kirkii*
Wildebeest *Connochaetes taurinus*	African Buffalo *Syncerus caffer*

A common and easily observed species is the graceful and extremely agile Impala, an animal of heavily bushed savanna and forest glades rather than open country. It is a beautiful antelope. The long and slender S-shaped horns of the male give the Impala its generic name *Aepyceros* – 'lyre horn'. The coat is glossy reddish-brown, and the buttocks carry a characteristic pattern of black and white, conspicuous as a signal to warn other members of the herd when one individual starts to flee from danger. At such times their movements are very fast and the

Defassa Waterbuck

animals make prodigious leaps, even when there are no obstacles to clear. Measured jumps 35 feet in length and 10 feet in height have been recorded; and 70 feet has been covered in three successive bounds of 26, 16 and 28 feet. These jumps enable the impala both to observe and confuse a pursuing enemy. From time to time other species such as Burchell's Zebra, gazelles, bushbuck or baboons join up with impala. Such an association provides for co-operation in the avoidance of predators and, as Schenkel reports, 'An alarm given by members of one species frequently alerts the whole association' (68).

Another antelope of woodland and clearings is the Waterbuck, a greyish-brown, shaggy-coated beast about the size of a donkey. The two species, Common and Defassa – easily distinguished by the rump pattern (page 174) – both occur in the Serengeti. Waterbuck stand with the head held high and have a dignified bearing; the males in particular, with their ridged, curved and widely-spreading horns, are magnificent.

They are appropriately named for, though not amphibious like the swamp-dwelling Lechwe, they do readily take to the water and are strong swimmers. At Paraa on several occasions waterbuck have been seen swimming across the Nile. When drinking they do not approach with caution and stand back from the brink like most antelopes, but wade right in up to their hocks. Consequently they are often taken by crocodiles, as I have found in Bangweulu Swamp and the Kafue River in Zambia, and below Kabalega Falls in Uganda. Under pressure from wild dogs they always make for the water, and have been known to submerge completely with only the nostrils above the surface.

The African Buffalo

Buffalo – 'the wild black cattle of Africa' – are found over a wide range of habitat. They are primarily grazers, but in forested areas they also browse on leaves and young shoots. Their main requirements are natural cover affording shelter in the midday heat, wallowing places, and an abundant supply of water. Like cattle, they must drink each day, usually in early morning and late evening, and this need limits their distribution. They enjoy wallowing; and resort to waterholes even when most of the water has evaporated – lying in the mire and emerging encrusted with mud.

In open country, buffalo become gregarious. Herds of thirty or forty head are common, and they are often found in herds of a hundred or more: in the Mara country herds of five and six hundred have been seen. Old bulls are rejected by the younger and more vigorous males and lead a solitary existence or congregate in small groups of outcasts.

On land their only enemy is the Lion. When a herd is threatened, the animals bunch together to form a defensive arc and can give a good account of themselves. But lone bulls are more vulnerable. According to Stevenson-Hamilton, buffalo are rarely killed by a single lion; but two lions will hunt together – one distracting the bull in front while the other attacks from behind endeavouring to hamstring him. Crocodiles prey on buffalo when they are watering, or, as shown in Plate 37, when they are crossing rivers.

The base of the horns forms a solid frontlet which, in an old bull, covers the whole of the upper part of the skull with a solid shield of armour. From the Treetops and other observatories where game come to water, under flood-light, one can see how effectively this shield is used at near ground level, when a buffalo is challenged by an aggressive

Buffalo at a water hole

rhino. In such places it is also evident that elephants will give ground to a herd bull.

When you approach buffalo, they stand and stare in suspicious curiosity, before turning to go off in a heavy lumbering gallop, only to stop for another truculent stare before again wheeling round. In a stampede the herd goes in a solid thundering mass.

When seen from a distance the animals might almost be mistaken for a herd of cattle. But at close quarters the illusion is dispelled. The thick-set body and muscular neck, the massive horns and drooping ears, the truculent pose as with muzzle lifted they gaze sullenly at the intruder – all seem to stamp the Buffalo as a dangerous adversary. And they are of course one of the hunters' 'Big Five' (the others are Lion, Leopard, Rhino and Elephant). A wounded buffalo has a well-deserved reputation for cunning and vindictiveness. Stevenson-Hamilton speaks of its determination to get even with the enemy. 'In fact a wounded buffalo will hunt his tormenter as a terrier does a rat.' Syd Downey regarded the buffalo as the most elemental of all the East African animals. 'They are endowed', he says, 'with ferocity and power and size . . . And once aroused . . . then the buffalo is formidable indeed – lustful for vengeance or destruction, insensitive to injury or obstacle, massively protected, with an appalling momentum' (17).

Wildebeest multitudes

The Wildebeest is a grotesque creature. The strange long black face with its Roman-nosed profile is over-large and gives the animal a doleful expression: with ox-like horns, a dark equine mane, a beard, drooping back, thin legs, and the tail and hind quarters of a pony, it is not surprising that Zulus formerly believed the species to be a hybrid – the progeny of buffalo and zebra.

In action he has the making of a clown. As you approach, he will stand stock still, staring at the intruder. Then with a twitch of his long tail, he springs round and gallops madly away, making occasional sudden swerves and capers, bucking and plunging with the head carried low and tail whisking wildly.

Wildebeest are highly gregarious creatures of the open plains, where they may be found in close association with Burchell's Zebra and gazelles. They are active at all hours of the day and night, and unlike most other antelopes they are noisy, uttering low resonant moans. The alarm signal is a series of hoarse snorts. In short grassland and open savanna the wildebeest is a key species, exclusively a grazer, and locally present in what seem to be incredible numbers. Its range extends from Kenya east of Lake Victoria southward to Natal and westward to Angola and South West Africa.

Unlike the gazelles, wildebeest are among the first open country

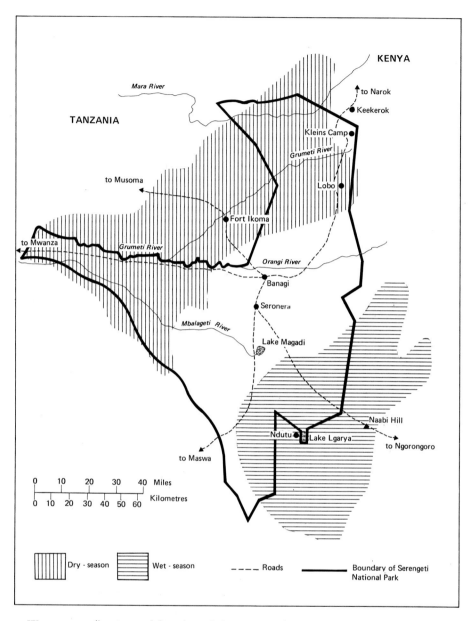

KENYA

Mara River

TANZANIA

to Narok

Keekerok

Kleins Camp

Grumeti River

to Musoma

Lobo

Fort Ikoma

to Mwanza

Grumeti River

Orangi River

Banagi

Seronera

Mbalageti River

Lake Magadi

Naabi Hill

Ndutu

Lake Lgarya

to Ngorongoro

to Maswa

| 0 | 10 | 20 | 30 | 40 | Miles |
| 0 | 10 | 20 | 30 | 40 | 50 | 60 | Kilometres |

Dry - season   Wet - season   ---- Roads   —— Boundary of Serengeti National Park

Wet season (horizontal lines) and dry season (vertical lines) distribution of migratory Wildebeest in the Serengeti area

antelopes to suffer from drought; and their extensive wanderings are largely to be explained by the availability of water and grazing. Much of their life is occupied with grass-seeking movements – the quest for pasture revived by unreliable and localized rainfall. But in time of

47

drought they are dependent upon permanent water supplies. Thus in the wet season huge nomadic herds are dispersed over the short grass areas, while in the dry season the large groups break into smaller herds concentrated in holding areas where perennial water is to be found.

In South West Africa wildebeest are unwilling to leave water-pits so long as, by scraping with their fore-feet, small puddles remain, and many of these pits are kept open, partly by wildebeest and partly by warthog, for weeks after they would otherwise have become dry. These isolated drinking places are made conspicuous in the early morning and evening when clouds of sand-grouse and doves collect at the water (70).

In the Serengeti the movements are regular, and the huge population migrates annually between the dispersal area in the short grass plains and the dry season concentration areas to the north-west. These migrating multitudes present a spectacle that is unique, and something to be wondered at. In 1967, when Murray Watson published the results of his three-year ecological study (82), the wildebeest population was estimated at 380,000 – by far the densest aggregation of large mammals to be found anywhere in Africa.

In the wet season – roughly December to May – the wildebeest graze in the open plains to the south east of the Park and beyond its boundary. During these months they congregate in vast nomadic herds, or massive 'armies', numbering up to 100,000 head. The rutting season is in late May and early June. Then commences the spectacular north-westerly migration. The animals tend to move in columns or in single-file. Seen from an aircraft they present an extraordinary spectacle, the advancing columns of blackish bodies calling to mind colonies of driver ants on the march. Clearly defined routes are used year after year, and in many places the trampling hoofs have cut deep grooves in the ground.

Once arrived at the dry season holding area the wildebeest split into progressively smaller concentrated groups. The return movement begins in November. Some early calves are dropped en route, but the main calving takes place after return to the short grass plains in January. At this time of year, and during the southward migration, predatory camp followers – lions, hyaenas and jackals on the ground and vultures and marabou storks in the air – are much in evidence, the smaller mammals waiting their chance to pick up afterbirths and dead or lost calves.

Very remarkable is the synchronization of calving, reported by Watson, which shortens the period during which young animals are most vulnerable to predation. 'Certainly the success of calving', he

1. Portrait of a Lion

2. Hippopotamus grazing by day. An unusual view of the animals on parade and dressed by the left

Courtship scene in the Kazinga Channel, Uganda. A Common Sandpiper is perched on the male's back

3. Black Rhinoceros with ectoparasites on its flanks, Amboseli

4. Bull Elephant walking in South Luangwa Game Reserve, Zambia

Bull Elephant demonstrating: a shake of the head has thrown up his left ear

5. Jackson's Hartebeest by the Albert Nile, showing extreme length of the head and frontal pedicel, and V-shaped horns

Chimney-like termite mound near Moroto in Karamoja, Uganda

Lion yawning beneath thorn acacia, Seronera

Black-backed Jackal, a constant camp-follower of the Lion

6. Bosc's Monitor (*Varanus exanthematicus*), an inhabitant of dry stony country, has bead-like scales and powerful claws. Karamoja

The Leopard Tortoise (*Testudo pardalis*) is the largest species in Africa: specimens weighing up to 75 lb have been recorded

7. Wildebeest, or White-bearded Gnu. Males are contesting territory

A nursery herd. Female Wildebeest with calves, Ngorongoro Crater

8. Male Olive Baboon, Nairobi National Park

Rock Hyrax. Despite its superficial resemblance to a large guineapig, this animal – the 'coney' of the Bible – is more nearly related to an elephant than to a rodent

9. Burchell's Zebra grazing near a stand of Fever Trees, Ngorongoro Crater

Part of a large herd of Topi at Ishasha in south-west Uganda

10. Masai Giraffe at the foot of the Rift Valley escarpment, Manyara

11. A fine Eland resting in Ngorongoro Crater. The rather short horns, present in both sexes, are keeled and spirally twisted

Common Waterbuck. The curved horns, carried only by the male, are strongly ringed

12. Female and young Defassa Waterbuck. The broad white rump-patch distinguishes this species from the Common Waterbuck

Pre-mating behaviour of Defassa Waterbuck. The long hair gives the coat a shaggy appearance

13. Siesta in the wild

Young bull Buffalo in the Kazinga Channel

14. Impala. Slender lyre-shaped horns are present in the male only. The rump-pattern of this graceful, high-leaping antelope – a diagonal black streak on the white buttock – is a distinctive character

Uganda Kob. The lyrate horns, also present only in the male, are robust; the body is thick-set and the coat reddish-brown

15. A fine black-maned Lion seen in profile, Seronera

Young Lion at rest, Ngorongoro Crater

16. Lion sheltering from the mid-day sun, Rwenzori Park

writes, 'depends to a large degree on the co-operation of all the animals of the group in which it takes place. During calving a number of cows lie down in a central area and as many as twenty calves may be produced in one hour. Surrounding cows become much more alert and individually chase jackal away from the area.' A number of cows have been seen to co-operate in driving single hyaenas from calving cows. Calves are precociously active: they are able to stand up four or five minutes after birth and can make their first uncertain steps five minutes later (82).

The Giraffe

The Giraffe, exclusively African and the tallest animal in the world, is an inhabitant of dry bush and acacia woodland. Being entirely a browser, it is not found in open grassland, or in thick forested country.

Systematists divide the species into a number of races or subspecies, distinguished by colour and pattern and horn formation; but intermediate stages between the different types are found. Two races of the Common Giraffe occur in East Africa: the Masai Giraffe, with jagged-edged or 'maple leaf' markings, in south-west Kenya and Tanzania; and the Baringo or Rothschild's Giraffe, with more rounded markings, and three or five horns, in western Kenya and northern Uganda. By far the most handsome is the Reticulated Giraffe, often regarded as a distinct species *Giraffa reticulata*, found in the Northern Frontier region of Kenya. It is a rich liver-red in colour with a net-work of narrow white lines.

Some people regard the giraffe as ungainly in appearance and awkward: certainly when seen in a zoo it looks an outlandish and incongruous creature. But in its native acacia parkland it is part of the landscape – with poise and a strange beauty all its own. And under observation it reveals itself as a creature wonderfully adapted to fill a special niche in the savanna community.

Sight of the giraffe is very highly developed, and is believed to be the best in all African game. The animal's height – as much as eighteen feet in a large male – gives command of a long field of view. This he maintains even when sitting down to rest, for he never lies flat on the ground but always keeps the neck erect.

Owing to its extreme length one might almost expect the neck to have more cervical vertebrae than, say, a whale or mole or man. In fact, the normal complement of seven is present; but these bones are greatly elongated, and unlike those of most mammals, they have a ball-and-socket articulation which gives the neck the extraordinary flexibility displayed when giraffes are 'necking'.

Reticulated Giraffe pacing

Special modifications of the circulatory system are related to the animal's height. The heart must exert great force to pump blood a distance of ten or twelve feet up to the head; and not surprisingly it is massive, with muscular walls three inches thick. The lower part of the very long carotid artery – the vessel that carries blood to the brain – has its walls thickened with elastic tissue to withstand blood pressure nearly three times that normal in man. The jugular vein which returns blood to the heart is enormous, being more than a inch in diameter. Its thick walls are furnished with numerous valves which prevent blood flowing back to the head when the neck is lowered.

At the base of the brain the carotid artery passes into a knot of fine vessels called the *rete mirabile* or 'extraordinary net'. This network supplies and regulates the flow of blood to the brain. When the giraffe lowers its head, for example in drinking, the vessels expand to accommodate excess of blood flowing in under pressure of gravity; and when the head is again raised the vessels of the net contract, thus countering the greatly reduced blood pressure and the danger of blackouts.

The gait, both in walking and galloping, is peculiar. Owing to the length of the stilt-like legs and proportionately short back, if a giraffe were to walk or trot like a horse it would trip itself – the hind-foot treading on the fore-foot. The walk is a slow rolling pace, the amble, in which both legs of the same side move together. It shares this gait, which

Masai Giraffe feeding on Whistling Thorn

is sometimes called 'pacing', with the Camel. In the gallop, both fore-limbs and both hind-limbs move together; and the hind-limbs are splayed widely apart so that the hind-feet can overreach the fore-feet by swinging outside them. Galloping giraffes seem not to travel fast, yet with huge strides they can cover the ground at 35 m.p.h. – going with a curious rocking motion and with the tail curled over the back.

They browse on the leaves and twigs of many trees such as *Acacia*, *Balanites* and *Scutia*, and in some areas are addicted to *A. mellifera* and *A. drepanolobium*. In Nairobi National Park they often stoop to feed from low bushes of the latter. In feeding, leaves are collected and stripped with the prehensile lips and extensible tongue which is about seventeen inches long.

Giraffes are most often encountered singly, or in small groups; but large herds of over fifty animals together have been recorded. Their only enemies, apart from man, are lions. Solitary bulls are not infrequently killed, and animals that have been surprised in woodland or when drinking. But in defence they are huge and formidable, the front legs with their massive hoofs being used as choppers.

Males fight a great deal, staging duels which are known as 'necking'. Rivals take up a position side by side, facing in the same or opposite directions, lean against each other with the legs straddled apart, and then use the head as a sledge hammer. In young animals the neck weaving and delivery of blows is half-hearted. But adult bulls swing the head backwards over the shoulder to strike the opponent's back, flank or belly with such force that the sound of the blows can be heard fifty yards away. The sparring may continue for twenty minutes, and other giraffes are sometimes attracted and stand watching.

Spinage has shown that the skull of the male becomes progressively reinforced, with age, by the deposition of bone, which adds to its weight, strength and effectiveness as a weapon highly specialized for this peculiar mode of intra-specific combat (72).

Maternal solicitude

There is no fixed breeding season, single young being born at any time of year after a gestation period of about 440 days. At birth the calf is already six feet tall, and weighs 120 lbs (35). According to Foster, the bond between mother and young is loose – perhaps more so than in any other species of ungulate. Most young remain with the parent for only six weeks: they can be weaned at a month, and have been seen browsing on whistling thorn in their first week of life (27).

17. Sunrise at Paraa, Kabalega Falls National Park

Uaso Nyiro River, Samburu Game Reserve

18. Hippopotamus and calf, Kazinga Channel, Uganda

Spotted Hyaena with cubs, Mara Masai Game Reserve

The brief association between mother and calf, and the tendency of the young to wander, has led many observers to believe that the maternal instinct of giraffes is poorly developed. But as Guggisberg has pointed out, the calf is able to remain in visual contact with its mother and to receive her visual signals over a considerable distance (33). It is also known that when a calf is in danger of attack, the mother will place it between her fore-legs and fight off assailants with smashing blows from her hoofs (74).

In the Serengeti, near Lobo, I once witnessed a profoundly moving demonstration of a giraffe's solicitude for her calf. Guided by vultures overhead and by the presence of two roving hyaenas and a lone giraffe we had come to a spot where there lay dead a baby giraffe. The animal had evidently just died and when first seen it was unmarked by any predator. Several White-backed Vultures flew down from the tree-tops. As soon as the first vulture tore at the carcass the female giraffe, who had been watching from nearby, strode up and stood over the calf. With her legs widely splayed she reared up twice, bringing her feet down with tremendous chopping blows. She then walked away a hundred yards and stood looking back in the direction of her calf. The vultures, now joined by a Ruppell's Griffon, began to gather round. Meanwhile one of the hyaenas approached, drove off the birds, and with difficulty began to drag the carcass away. By exerting his utmost strength he had half pulled and half carried the calf about twenty yards when the mother returned: the hyaena slunk off at a canter. We gazed in wonder as the startling drama unfolded.

The giraffe now stood in front of her calf, and with her neck bent low she regarded it closely. Then taking a pace forward she again straddled the calf, jumped and brought down the front legs, held stiffly like ram-rods, to strike the ground with terrific force. Using her widely-splayed front legs like pile drivers she repeated the chopping kick, and then remained standing over the baby, with her neck arched over and a long thick rope of saliva dangling from her mouth. She then retired; but as soon as the vultures began to collect she returned yet again to drive them away. She then slowly lowered her head to smell and gaze at her calf. Once more she took up her stance above the body and executed the sudden steam-hammer blow, shaking the ground with the thud of hoofs. After a pause she walked away and stood again to keep watch from nearby. The hyaena reappeared – not to feed *in situ* but to drag the crumpled corpse away. We did not stay to watch the tragedy played out. From a scene of such impressive significance one could but drive away in subdued and silent mood, the heart disturbed by feelings of awe and sadness.

Grant's Gazelle

Thomson's Gazelle and fawn

Niche structure of the ungulates

No one can visit the game areas of East Africa without being impressed, and indeed astonished, by the abundance and rich diversity of ungulate species. 'This variety does not exist for our amusement or for our contempt', wrote Fraser Darling in his essay on the ecology of biological communities in Zambia, 'it has evolved as a complex of creatures making the fullest possible utilization of part of a habitat. A characteristic phenomenon in evolution is that species differentiate to states in which they overlap others as little as possible'(19).

Ecological separation of ungulate species living in close contact may be achieved in several ways (43). In the first place, species occupy different habitats. Thus Wildebeest, Zebra and Grant's Gazelle prefer open grass plains; Oryx and Gerenuk are tolerant of arid conditions; Dik-dik and Lesser Kudu live in densely wooded areas.

Again, species differ in their feeding habits. Topi, hartebeest, wildebeest, buffalo and zebra are either wholly or primarily grazers; others like Black Rhinoceros, Giraffe, Eland and Bushbuck are browsers. Others again – browser-grazers – have intermediate habits: examples are Waterbuck, Reedbuck, Bushbuck and Lesser Kudu. Elephant and Warthog are exploratory general feeders, taking a wide spectrum of plant products.

Species living in the same habitat may be ecologically separated by divergent food-preferences. For example, observations by Leuthold have shown that Gerenuk feed almost exclusively on leaves, shoots, flowers and a few fruits of trees and shrubs, but not at all on grass and small herbs. In this respect it differs from the Lesser Kudu which eat some grass, and a variety of annual and perennial herbs (47).

Migration is another factor in segregating species. For example, zebra, wildebeest and buffalo move into the concentration areas where there is permanent water in the dry season, while Grant's Gazelle, which can live long without drinking, make use of the Masai steppe. An interesting example is described by Vesey-FitzGerald from the Athi-Kapiti plains of Kenya, where it was found that the movement of gazelles tended to alternate with that of the large ungulates. The gazelles remained in the wet-season dispersal areas during the dry season when zebra and wildebeest had moved to the water-holding concentration areas, but migrated into these areas when the rains commenced, to benefit from the short-grazed grasses left by the heavier game which had meanwhile returned to the dispersal areas (79).

Browsing animals feed from different strata of the vegetation. The

Gerenuk, or Waller's Gazelle

Giraffe can crop acacia leaves to a height of seventeen feet. The Gerenuk, though a small gazelle, can reach up to a height of seven feet – thanks to its long neck and habit of standing erect on its hind legs. The Black Rhinoceros browses at a lower level, often from the gall acacia or

Whistling Thorn. Dik-dik browse near to the ground and can enter the dense canopy of shrubs for food that rhinos do not reach.

In some cases, as Field (25) has shown in Uganda, different grazing herbivores may utilize different parts of the same plant species. Thus antelopes eat the leaves, buffalo the leaves and stems, and warthogs the storage bases of the grass *Sporobolus pyramidalis*.

Resistance to drought

A remarkable phenomenon is the capacity of some ungulates to go waterless. The experienced hunter and naturalist T. R. H. Owen, who was for nearly thirty years a Political Officer in the Sudan and later served as Deputy Game Warden in Uganda, considered it 'one of the greatest and most inexplicable wonders of the animal world' (61).

Game differ enormously in this respect. Some, like the Elephant, African Buffalo, Waterbuck, Uganda Kob, Roan, and Zebra, are very susceptible to drought: they are confirmed drinkers and are never found far from water. Others, for example Giraffe, Kudu and Beisa Oryx, can dispense with water for long periods, but nevertheless drink when it is available. Eland drink regularly in country where water is plentiful, yet they can resist drought. Dik-dik inhabiting waterless regions survive without drinking. Gerenuk are said not to drink at any time. In true desert animals – Dorcas, Addra Gazelles and Addax – the extraordinary capacity to live waterless is universal 'and that under conditions of temperature and humidity,' as Owen says, 'where a human could die of thirst within twenty-four hours and could not survive forty-eight.' Strangely enough these specialists, physiologically adapted to resist desiccation, do not drink even when water is available: they are total abstainers.

Water independence is also found among the predators. Owen believed that some jackals and desert foxes never see water. On several occasions Downey came across Kenya Lynx right out in the desert of the Northern Frontier where the animals could get no liquid except blood (17). Lions will penetrate into waterless areas. Formerly they were found in the Red Sea Hills of the Sudan. Early in the century they were reported to be numerous in waterless north-west Kordofan. Brocklehurst (*Game Animals of the Sudan*) found remains of an oryx which had been killed north of latitude 16° in country where the only water was in isolated wells thirty to forty feet deep. 'The spoor of this particular lion went away from the kill as straight as an arrow into the desert, but owing to insufficient water I was unable to follow it.'

3
The Carnivora

And, above all others, we should protect and hold sacred those types,
Nature's masterpieces, which are first singled out for destruction on
account of their size, or splendour, or rarity, and that false detest-
able glory which is accorded to their most successful slayers.

W. H. Hudson

THE African fauna is richly endowed with members of the Carnivora.
They range from the lion – master predator who can kill an adult
buffalo bull or male giraffe – to very small forms like the Dwarf Mongoose whose food is mainly insects.

Carnivora in East Africa

These animals may be conveniently divided according to size into three
categories. The first contains the three outstanding species – Lion,
Leopard and Cheetah. The second – Wild Dog, three species of hyaena,
of which the Spotted Hyaena is the most common and widely distributed, and the Caracal. The smaller members, belonging to several
families, are of course the most numerous and include three species of
jackal, the Bat-eared Fox, Honey Badger, weasels, otters, civets, genets,
many species of mongoose, the Serval, and other small cats.

The striking fact about the East African species is their number and
diversity. This will be appreciated by referring to the list of species which,
for example, have been recorded from a single National Park – the
Serengeti of Tanzania.

CARNIVORA

CANIDAE

Golden Jackal *Canis aureus*	Bat-eared Fox *Otocyon megalotis*
Side-striped Jackal *C. adustus*	Wild Dog *Lycaon pictus*
Black-backed Jackal *C. mesomelas*	

MUSTELIDAE

Zorilla *Ictonyx striatus*	Honey Badger *Mellivora capensis*
African Striped Weasel *Poecilogale albinucha*	Cape Clawless Otter *Aonyx capensis*

58

Each of these species occupies a particular niche, and plays a more or less distinct role in the ecosystem, the various members of the order making, between them, the fullest utilization of the terrain, with its food resources, that is their home.

In the following short account we shall be looking at some of the larger, more spectacular and better known felids and canids.

The Lion

The Lion is distinguished from all other cats by the massive mantle of hair carried on the neck and shoulders of most adult males; by the black

Lions near Lake Lagarya

terminal tail tuft, with its curious concealed horny spur; by its social habits and by its voice.

The roaring of lions is the most remarkable and moving sound of the African bush. Heard at night from camp it never fails in its power to stir the imagination. In its urgent quality and tremendous volume there is no other sound in the least like it. Colonel Stevenson-Hamilton has compared the roar, heard at close quarters, to 'the rumbling of an immense bass organ ... the great volume of sound dying away in a series of *diminuendo* grunts'. On a still night lions can be heard over a great distance – perhaps up to five miles. To hear members of a pride thus communicating over long distances is a thrilling wonder.

The strong development of sociable behaviour is another feature unique among the true cats. Lionesses are rarely seen alone. Members of the family-group stay together, and their numbers continue to grow. Very large groups are sometimes seen, composed of one or more adult males, several females and – the majority – of cubs and adolescents. The largest pride I have encountered was in October 1971, near Seronera, when one hot afternoon we came upon a party of ten lionesses and twenty-four younger animals of various ages sprawled together under the shade of a wide-spreading tree. It was an astonishing sight – the ground carpeted with the prone, dun-coloured forms. A hundred yards away two large males were taking their siesta in a thicket apart.

Prey of the Lion

The Lion's prey includes the larger and medium-sized ungulates, such as Zebra, Wildebeest, Hartebeest, Topi, Impala, Eland and Buffalo; also young of elephant and hippopotamus. Warthogs are sometimes dug from the burrows they occupy at night. In some areas full-grown giraffe and buffalo are pulled down. Lions will also take baboons, monkeys, poultry and even fish, pythons and crocodiles.

Crocodiles, with their extremely keen olfactory sense, are quick to take advantage of another animal's kill, and there are reliable records from Tanzania and South Africa of lions killing a crocodile in such circumstances. Lions will sometimes deliberately hunt the reptiles. At Butiaba in Uganda a lion killed a crocodile measuring eleven feet seven inches and consumed the neck, shoulder and flanks. Pitman reported in 1942 that one afternoon visitors to the Kabalega Falls saw a lion stalk and kill a crocodile on the opposite bank: the creature was so hungry that it continued to feed even when the launch crossed the river to afford a closer view. Lions have also been known to kill and devour crocodiles

19. Grevy's Zebra, Northern Province, Kenya

Cape Buffalo bull, Lake Manyara, Tanzania

20. Rock Hyrax, Serengeti National Park

Unstriped Ground Squirrel, Samburu Game Reserve

on the Lake Mobutu flats; and this is regarded as quite a normal occurrence on the western shore of Lake Rudolf.

In some areas the choice of prey seems to be influenced by the density of prey species. For instance, in their study of lion predation at Lake Manyara, Makacha and Schaller (50) found that of 60 prey killed, 37 (62 per cent) were buffalo, the most abundant large mammal in the Park.

But choice is not determined solely by relative abundance – even between prey of about the same size. In Nairobi National Park Professor Bourlière (3) found that about 50 per cent of kills were wildebeest and only 15 per cent were zebra; whereas their relative abundance was 35 per cent and 24 per cent respectively; and the number of Kongoni killed (2 per cent of the total) was proportionately far below their availability (15 per cent) as potential prey.

Hunting methods

Lions hunt by sight and sound rather than by scent. Hunting is mainly a nocturnal occupation, but lions move about freely in the early morning and evening and in country where they are not themselves hunted, as in National Parks, diurnal hunting is often seen.

The methods adopted depend upon the terrain and the quarry. Lions are intelligent animals, and the hunting pride is a well-organized group whose individuals understand the use of ground, cover and wind-direction. When game is heard or seen several lionesses stalk off and hide down-wind from the quarry. Once they are in position others of the pride circle round up-wind and then, with scent streaming down towards the prey, they act as 'beaters' and stampede the herd into the ambush. Usually it is the females who do the actual killing.

Often lions rely on the stalk-charge technique. One or more individuals make a slow, concealed approach, which may last for hours, until they are near enough to charge down upon the quarry. Col Stevenson-Hamilton (74) said that a herd of game may be entirely surrounded by a large pride of lions whose approach, owing to lack of wind, had not been detected.

Killing is usually done by strangulation. Often the fleeing animal is thrown off balance by the final charge and breaks its neck in falling. Large prey such as giraffe and buffalo are killed by the males of the pride, and then sometimes only after a prolonged struggle during which the quarry may first be ham-strung and thus disabled. Buffalo in a herd combine in defence against lions, facing the enemy in a compact armoured front. Solitary individuals – usually males – lack such protection, and Makacha and Schaller found that thirty of the thirty-seven buffalo kills they examined were males. Herds of the Manyara buffalo sometimes succeeded in driving off the attacking lions. In one case, after killing a buffalo, five lions – a male and four females – had to take refuge in a tree from the herd two hundred strong, and were unable to come down to their meal until the herd finally left in the evening. On another occasion Park rangers saw a herd chase a lioness and three cubs. Two cubs and their mother escaped by climbing: the third cub was trampled and killed (50).

Significance of sociality in the Lion

The importance of sociality in the Lion has been stressed by Sir Frank Fraser Darling (18). The loose structure of the pride, as he points out, is of the highest survival value to a species that, on account of its size, has got into an evolutionary cul-de-sac.

The cubs are born small and helpless; and since the coverts where they lie are accessible to jackals and hyaenas, they cannot be left alone. Danger from predators is averted by some lionesses of the pride staying with the cubs while others hunt. Litters of about equal age are nursed promiscuously by females of the pride. Unlike other cats, lions do not bring meat to the cubs, so they depend entirely on the mothers' milk until they are old enough to walk to a kill.

So it is that the questionable asset of size has been met by sociality and division of labour, both in hunting and in rearing offspring. From his ecological studies in the Mara plains of Kenya Darling concluded that if lions become so reduced in an area that the females cannot combine, they may be assumed to be finished as a successful breeding species.

The Leopard

Unlike the Lion, which is an inhabitant of lightly wooded country, open grassland, or even sub-desert, the Leopard's preferred habitat is where cover abounds, in gallery forest, dense bush, thickets and rocky ravines.

Leopards are unsociable, and they are generally silent. The voice, once heard, is unmistakable – the call consisting of harsh coughing grunts repeated in quick succession, somewhat resembling the sound made in sawing wood. Except during mating time the male and female lead solitary lives.

Sleeping Leopard

In temperament the Leopard is most unpredictable, and it has a reputation for stealth, daring, and intrinsic savagery. This side of his nature is illustrated in Plate 50. The animal was surprised in long grass: his jaws are parted; the upper lip is inflated and drawn back in a snarl which pulls the skin into wrinkles about the nose to show the long white

63

canine teeth; and above all the eyes, fixed in a straight steady stare at his enemy, seem filled with murderous intent.

Prey and hunting behaviour

Leopards are nocturnal hunters, and have been said to hunt entirely by night. But wild animals follow no fixed rule. We are often told that a species behaves in this way, or in that way: yet the fact is that individuals differ greatly from one another; and an individual adapts its actions to circumstances.

An incident I shall not forget occurred one morning in 1962, at about midday, as I was driving to the Kabalega Falls from Masindi. A family party of warthog – two adults and four young – scampered on to the road from the long grass verge. They were in disorder and appeared bewildered. One adult hesitated, and rushed back into the grass, when suddenly from the same side a leopard sprang on to the road. Amidst a flurry of dust and squeals of alarm, the big cat seized a piglet, carried it across the road and climbed with it into a tree. The whole action was over in a few seconds.

Leopards prey on a variety of small game. In the Serengeti Thomson's Gazelle are the most important prey; in the Kruger National Park, Impala. In addition to other small antelopes, hares, cane rats, and hyrax, the Leopard takes game birds and domestic poultry. It is partial to Bush Pig. Baboons are another favourite prey, and the present-day scarcity of Leopards may in-part account for the large population of baboons which are so destructive of crocodile clutches in Uganda. There are many accounts of leopards entering houses at night and taking dogs – even from off a bed where its master is sleeping. An incident of this kind is described by Temple-Perkins: 'I have known one come into a bedroom occupied by two brothers, both light sleepers and first-rate shikaris, and take a dog from a mat between their beds without either of them waking' (77).

The hunting technique of the Leopard and of the Lion differ in many respects. The Leopard is a solitary hunter, and it is unusual to find more than one on a kill. In their fascinating study of predation in the Serengeti area, Kruuk and Turner (40) describe how a Leopard, when moving in for a kill, takes full advantage of its covered habitat. It stalks with the body held touching the ground, 'proceeding with almost snake-like movements, eyes and ears fixed on the spot where the prey is'. From the stalk-ambush position the fast final dash is very short: usually the prey can be reached in one bound. The prey is taken to a tree, and hung over a horizontal limb or wedged into a fork, where it is secure from the attentions of lions, jackals and hyaenas.

The Leopard's larder

One day a safari group visiting Seronera had a wonderful demonstration of the use of a tree larder. We had come upon a leopard cub perched in an *Acacia tortilis*. Two hundred yards away, and eighteen feet up in another acacia, a full grown and freshly-killed Thomson's Gazelle was draped over a branch; the female leopard lay sprawled nearby. As we watched she climbed to the kill and began to prepare the meal. First she scraped fur from the flank and spat it out. Then, opening the abdomen, she tore out and partly ate the intestines while rejecting their contents. The stomach was similarly treated. She then jumped down to a lower branch and called twice to her cub, who trotted over, nimbly climbed into position, and began a prolonged meal – eating from the kidneys, liver, heart and lungs. This done, the cub jumped to a more comfortable seat, washed face and forepaws like a domestic cat, finally climbing to his mother's side to be licked clean. The Leopard's arboreal habits enable such domestic matters to be settled without haste and free from disturbance.

Even so, the larders are sometimes raided. A year ago near Banagi in the Serengeti we saw a troop of baboons playing beneath a tree in which the remains of a gazelle were suspended. Two members of the party were up in the tree, eating from the carcass. Needless to say – the leopard was not at home.

Cheetah – greyhound of the Felidae

The graceful Cheetah is an aristocrat of the animal world. It is also the fastest animal on earth. Long in limb and slender in build, it is entirely fashioned for speed – a greyhound among the Felidae. Its slim body, small rounded head, and long legs distinguish it at once from the Leopard. The round black spots of the coat are single as in the Serval, not in groups of five as in the Leopard. A characteristic feature is the black stripe running from the inner-corner of the eye to the angle of the mouth. A slight ruff on the neck extends past the withers, and the bushy tip of the long tail is white. Unlike the true cats, it lacks fully retractile claws.

The Cheetah is less nocturnal than other members of the family, and much of the hunting is done by day, mostly in early morning and

Cheetah looking back

late afternoon. It preys mainly on smaller antelopes up to the size of
Grant's Gazelle or Impala, and on hares, bustards, francolin and guineafowl. In a detailed study of the Cheetah's hunting behaviour in the
Serengeti, Schaller (67) discovered that 89 per cent of kills were Thomson's Gazelle – the most abundant species of appropriate size in the Park.

Although Cheetah and Leopard tend to feed upon the same species, they are ecologically separated – the former hunting in open plains and the latter in thick cover.

The Cheetah's method of hunting is quite different from that of other cats. Relying on its ability to outstrip the fastest antelope over a short distance, it selects its victim and rushes down upon it with burning speed, knocking its hind feet from under it with a fore-paw and seizing the throat to kill by strangulation.

Being unable to defend its kill, or to climb with it, a cheetah settles down to eat its fill hurriedly, before vultures have revealed the position of the carcass to lions and their camp followers, the hyaenas and jackals.

The Cheetah is a gentle, timid and inoffensive animal, docile and responsive in captivity. Vulnerable to changing conditions in Africa and sensitive to increased disturbance – and in this respect tourism is not blameless – this sprinting world record-holder is fast losing ground in the race for survival. Cheetahs have been exterminated in southern and north Africa. Elsewhere their range is shrinking and their numbers dwindle. 'In Kenya,' wrote Noel Simon in 1962, 'the species is balanced precariously on the brink of extinction and may already have passed the point of no return' (71).

Family of four

The Spotted Hyaena

Of the three African species, the Spotted Hyaena is the largest, commonest, and most widely spread, ranging as it does almost over the whole continent south of the Sahara, from sea level to the snow-line of Kilimanjaro, and from semi-desert to forest.

In appearance hyaenas are not attractive, and they are often referred to by such adjectives as 'uncouth' or 'repulsive'. Sir Samuel Baker branded them as 'low-caste creatures'. From the massive head and ruffed neck the back droops away behind, so that the fore and hind quarters

Hyaenas at crocodile's nest, Namsika, Victoria Nile

seem out of proportion; and when on the move the animal has a slink-ing lope. The furtive, cringing manner is also deceptive, for the hyaena is both cowardly and ferocious. Where bold and numerous, as formerly in north-west Botswana, there were reports of hyaenas severely mauling sleeping natives and dragging children out of the huts at night. In the 1950s in Nyasaland there was an outbreak of killings, mostly in hot weather when people were sleeping out. Theodore Roosevelt believed that they were far more apt than Leopard to prey upon human beings. 'Their attacks are always made at night, with extreme caution, and when the victim is sleeping. Usually they seize the face . . .' (66).

The calling of hyaenas is one of the familiar sounds of the African night. The typical cry 'whoo-eee-ooo' begins on a low, hoarse note, rises in a slurred interval of a musical fifth, and then drops back to the first tone. This eerie call is only one of a variety of sounds and modu-lations of voice by which the animals communicate over long distances. The most hideous concerts and ghoulish chuckles are heard when hyaenas are stealing a kill, or mating. Dr Harrison Matthews has de-scribed the scene: 'Shrill shrieks and yells, accompanied by deep emetic gurgling, made a background for peals of maniacal laughter.' When an over-venturesome hyaena had been wounded by a lion at a kill, 'its last cries were accompanied by peals of laughter from the other hyaenas waiting in the background' (52). Such music is not to everyone's taste; and it is not surprising that the hyaenas play a part connected with witchcraft.

When hunting by day hyaenas share with lions and jackals the habit of following vultures to the place where they congregate. In quest of carrion they will also follow lions and wild dogs. In the daytime, single hyaenas will pull down unprotected wildebeest calves. Near Seronera I have seen a hyaena hunt a young Thomson's Gazelle despite inter-vention by several adults which continually tried to head it off. One of the gazelles jumped clean over the hyaena as it relentlessly followed every erratic movement of the fawn. The end was inevitable – the prey was grabbed at the groin and carried away. At times they are surprisingly bold and will attack and devour a lion that is too old to defend itself properly. Occasionally they will follow a scent with nose to the ground, like a fox-hound; and in Rwenzori National Park I once saw a hyaena trail a fresh-water terrapin (*Pelusios*) which it seized and carried off. The fact is that in their feeding habits hyaenas are extraordinarily versatile opportunists.

Hyaenas' jaws are most powerful. On each side, above and below, there is a shearing 'carnassial' tooth, with a sharp V-shaped cutting edge.

Night scene on Lake George flats

These carnassials meet like bone-forceps, and the animal is even able to crack the femur of a hippopotamus, to get at the marrow. Bones and marrow form a considerable part of the diet. The droppings are consequently rich in mineral matter, and they dry out pure white. The animals use regular latrines, made conspicuous by the accumulation of white faeces.

It has generally been thought that the hyaenas' chief role was that of scavengers. But Dr Kruuk's remarkable observations in Ngorongoro Crater have thrown new light on the animals' strange and hitherto unsuspected night-life.

The mammalian fauna of the caldera is more or less sealed off from the surrounding country by high (2000 foot) walls. By marking – for subsequent identification – a known proportion of the hyaena population, Kruuk found that about 420 hyaenas inhabited the crater floor, and that this area (about 120 square miles) was divided into eight home-ranges – each range being the hunting ground of a 'clan' of from 10 to 100 hyaenas. The clans prey at night on wildebeest, zebra and Thomson's Gazelle. Of 1052 hyaenas which Kruuk observed feeding at night, 82 per cent were eating from an animal killed by hyaenas. It is clear that they are the key predators in the area. There is thus the bizarre situation that the normal lion-hyaena relationship is reversed: the Ngorongoro lions steal the kills and have become the scavengers. 'I have good evidence,' Kruuk says, 'that the lion population in the Ngorongoro Crater obtains its food largely in this way' (38).

On the boundless Serengeti plains Kruuk found that while some animals adhered to the clan system, many hyaenas were migratory and followed the wildebeest herds; while in a third category were 'commuters' that made long excursions, often lasting several days, between their permanent dens and the area where the herds happened to be grazing.

Hunting dogs

Wild Dogs – the wolves of Africa – are highly organized as team hunters. Nomadic in habit, they roam over the plains and only stay in one area when the whelps are too young to travel. Hunting is confined to the first and last hour of daylight. Otherwise daytime is spent by the pack members lying-up close together in cover. The packs number from six to thirty or more individuals: packs of over ninety have been recorded.

Antelopes ranging from small duikers to adult waterbuck and wildebeest are hunted. In Ngorongoro Crater and the Serengeti steppe Thomson's Gazelle are the main quarry. In other areas Impala and Reedbuck are favourite prey. The method of hunting has been described by Kruuk and Turner (40). The pack detect prey, by sight, at about 250 to 500 yards, and approach at a slow walk, only beginning to run when the herd wheel round in flight. 'Often several dogs would each chase an animal, usually shifting from one gazelle to another; soon one of the dogs, frequently the dog which had been leading the initial search, would go in pursuit of one individual gazelle and stay with that one. Immediately the other dogs then followed the leader.'

The trotting speed is about 7 m.p.h. and the sustained hunting speed 30 m.p.h. Contrary to stories that the prey-animal, once selected by the pack, is irretrievably doomed, the van Lawick-Goodalls found that only 44 per cent of 91 observed chases were successful (44). When coming in reach of the quarry, they bite where they can, causing it to fall, and then disembowel and tear it apart at once. Wild Dogs kill only what they need, and nothing is wasted.

Older members of the pack will give way to young dogs that are learning to kill. One evening in 1965 near Lake Lagarya a pack was seen to come upon a newly-born wildebeest that had lost its mother. The older dogs ran past, and left it to the younger members of the pack to kill their helpless victim.

The community feeding arrangements, as revealed by field observations of Kühme (41) and Kruuk and Turner (40) in the Serengeti, and Estes and Goddard (24) in Ngorongoro Crater are of great interest. In marked contrast to the behaviour of lions at a kill, adult wild dogs give precedence to young animals, standing aside until they have fed. The dogs then move in, bolt down hand-size pieces of meat without chewing, and run off to the den where they disgorge meat to the adults of both sexes that have stayed to guard the pups. For callow young the females bite up and later disgorge partly digested meat. If a mother dies adult males will continue to feed whelps until they are old enough to move with the pack. Adult dogs will also beg and receive food from their companions. If a lame animal reaches the kill late, to find everything eaten, it is fed by regurgitation as in parental feeding.

As a result of communal treatment of the food supply, 'the whole pack can be sustained on the produce of a few successful hunters'; and division of labour is possible between dogs that catch prey, adults that defend the family, and members of the pack that bring meat to the den (41).

Smaller Carnivora

In the short account which follows it is only possible to mention a few of the smaller species. The jackals are popularly regarded as scavengers, feeding on carcasses of animals killed by lion. But the Black-backed and the Side-striped Jackal, besides feeding on reptiles and insects, also kill for themselves small antelopes such as Dik-dik, duikers and Thomson's Gazelle. Estes (23) has described how both species employ the same efficient system of pair-hunting for gazelle fawns: one jackal concentrates on finding and catching the fawn while the gazelle is occupied chasing the other away.

When a lion pride is moving up for a hunt, Black-backed Jackals can often be seen lurking nearby, alert and taking a watchful interest in developments. At the kill they keep their distance until the lions have moved away, and then take their chance with vultures and hyaenas to snatch what they can.

It is an undoubted fact that when searching for a carcass jackals will watch the movements of vultures in the sky, and when they see them descending they hurry off towards the line of drop. Colonel Meinertz-hagen has described how on the Athi plains of Kenya he once came across a jackal actually lying on its back and turning his head in all directions as though watching something in the sky. 'Quite suddenly he jumped to his feet and scampered off, continually looking up into the skies, and on following his gaze I saw vultures descending from all directions on to a carcass . . . The vultures got there first but were soon scattered by the jackal and had to sit around while he had his meal' (53).

The Ratel or Honey-Badger is renowned both for its commensal relationship with the Greater Honey Guide (*Indicator indicator*) which feeds on bee-larvae after its partner has broken down the hive with its long claws in search of honey, and also for its fighting spirit, tenacity of purpose and utterly fearless temperament. And – a point to which I have referred on page 179 – its colouring is aposematic, as befits an animal whose strength, impervious skin, and power of endurance make it virtually 'indestructable'.

Dr Sylvia Sykes (76), who kept one of these animals in captivity, mentions what she calls the 'fury mood' when the Ratel reached a state of intense blind ferocity. In fighting with a dog the Ratel having once taken hold tenaciously retains its grip, allowing itself to be thrown about – jerking and twisting inside its tough loose skin – until the dog collapses from exhaustion. In this way the Ratel can overpower animals much larger than itself; and it is held in respect even by lion and leopard.

A ranger in Kruger National Park once saw a Ratel engaged in mortal combat with a ten to eleven foot python. Stones and dust were being scattered in all directions, and the fight continued for over fifteen minutes. When the ranger finally approached, the python was found to be 'so mutilated that it looked as if it had been run over by a train'.

In hunting mammals such as hares and small antelopes, ratels are said to trot unhurriedly with a long, swinging stride, overtaking their prey not by speed, but by sheer endurance; and according to Sykes, a

Portrait of a Serval

Ratel may cover twenty miles in a night's foraging. They are also said to assist each other by hunting in pairs; and in my experience in Uganda they do in fact sometimes work in pairs when excavating crocodile eggs at night.

The East African mongooses range in size from the Large Grey Mongoose, measuring 48 inches and weighing up to 8 lb, to the Dwarf Mongoose, a mere 12 inches long and a little over 1 lb in weight.

Spotted cat: the Serval

The larger species – Marsh, Large Grey and White-tailed – are nocturnal, and are usually to be found singly or in pairs. In marked contrast, the two small mongooses – Dwarf and Banded – are sociable, gregarious, strictly diurnal in their feeding habits, and take refuge for the night in warrens or deep chambers excavated in disused termite mounds.

One evening recently at Kilaguni I saw a pack of the Dwarf Mongoose returning from a foraging excursion. All seemed in a hurry to get home before darkness fell, and scampered in ones and twos or groups, all coming from the same direction. This particular pack had thirty-one members. At times these very gregarious animals travel in tandem so closely together that the pack in its winding course resembles a large fast-moving snake.

We have little detailed knowledge of the behaviour and ecology of the various species. But an exception is the Banded Mongoose which has been the subject of a fascinating study by Ernest Neal in Uganda (59). He found that dens in termite mounds, excavated by the mongooses

Bat-eared Foxes

themselves, had a central chamber some 60 by 35 inches in area and 20 inches high in the centre, with three or more entrance tunnels. The packs forage gregariously by day, investigating and tearing open elephant droppings in search of insects. Field observations and dung-analysis showed that the basic food animals were beetles – including dung beetles – and millipedes; and Neal draws attention to the interesting fact that this small mongoose is very dependent upon buffalo and elephant whose droppings provide its invertebrate prey.

4

The Four Giants

Africa is the land of wild beasts. The grandest forms of the terrestrial
creation have their habitation in that Continent.   *P. H. Gosse*

THE animals we shall briefly consider under this heading, taking sheer
bulk as the criterion, are the African Elephant (*Loxodonta africana*), Black
(*Diceros bicornis*) and White Rhinoceros (*Ceratotherium simum*), and Hippopotamus (*Hippopotamus amphibius*). The first and the last are important 'key'
species in habitats where, as in the National Parks of Uganda and Zambia,
they have become superabundant. The two rhinoceros species, on the
other hand, have been much persecuted and their numbers are declin-
ing.

Nowhere in the world outside Africa is a comparable mammalian
quartet to be found: but in the Kabalega National Park of Uganda one
may see all four species in a morning.

King of beasts

The African Elephant is the true King of Beasts. Among all the varied
assemblage of animals in Africa, he is the most imposing – the absolute
master. His huge bulk and unchallenged strength are tempered with
gentleness and sagacity. He has style, intellect, a natural dignity, a
majestic presence. If I may quote what I have said elsewhere: 'To walk
alone in the wild solitudes where elephants have their home is to feel
an indefinable thrill of wonder, known only to those who have experienced it. If you cherish ideas of man's superiority over the animal
creation, they will avail you nothing as you stand in the midst of a herd
of elephants. Your status is diminished and you meet the animals on
their own terms: you become aware that you are an intruder into a
world that rightly belongs to others' (13).

A fine trait in the elephant's character is the concern they show for
companions in distress, for the young, and the very old. The literature
contains many accounts of elephants coming to the assistance of a
wounded comrade and trying to help it to move out of danger. There is

Elephant at water

a report from Zambia of elephants coming back to the dead body of a companion that had been shot: they succeeded in lifting the head and front legs and persisted in their efforts for an hour and a half.

Very old animals carrying heavy ivory are always accompanied by two or three younger bulls. E. A. Temple-Perkins, one of the most experienced hunters, states in his fascinating *Kingdom of the Elephant* that he had *never* seen a really big elephant alone. At Ishasha in Uganda there lived until recently an old patriarch of exceptional size. On the occasion when Captain Frank Poppleton showed me this elephant in December 1962 it had in attèndance a very large bull and two younger bulls that appeared to be acting as pickets. The large guardian bull came up close, and is the animal I have drawn on page 207; but the old patriarch kept his distance and only gave us a glimpse of his massive tusks.

When a cow calves, she is closely attended by 'aunties' who leave the herd and play the role of protector and midwife. At Nyamasagani in Rwenzori National Park, Poppleton came upon such a scene. 'I heard a great commotion in the bush,' he writes, 'and found that it was caused by a herd of elephants. Part of the herd was spread out, but on the left there was a closely packed mass of animals all facing outwards. In the centre of them was a small black slimy object, which turned out to be a new born calf. The birth had clearly just taken place because the mother, easy to identify by her condition, and another cow elephant were removing the membrane covering the calf... The maternity group consisted of six adult cows and five young calves, with a young bull elephant looking on about fifteen yards away. Some of them guarded and assisted the new born baby, nudging and pushing the little fellow, using their trunks and feet to try to get him up. Others took the membrane bag, which had been removed from the calf, and threw it into the air so that it spread out like a blanket... Two hours after birth the baby elephant took its first stumbling steps' (63). When the baby is able to walk with the herd, its mother and other adults always shield it from harm and, when necessary, place themselves between the young one and danger.

The Lord Mayor

In temperament individual elephants differ, as do men. Sometimes they will destroy objects holding human scent; and their actions are unpredictable. There is a record of elephants burying a car beneath branches torn from trees. One afternoon in 1952 from my camp at Fajao below Kabalega gorge we heard trumpeting across the river. Presently a herd of about eighteen, including young of different ages, came through the bush on to a sandspit on which stood a grass-and-stick hide I was using to observe crocodiles. Three elephants detached themselves from the party, walked up to the hide and stood round regarding it. Then a large male stepped forward, his ears extended, and demolished the hide with a swipe of his trunk.

In the early days at Paraa, before the present Lodge was built, a bull elephant that became known as the 'Lord Mayor' frequented the site.

It was not unusual for this animal to remove thatch from the food-store, pry round the interior with his trunk, and help himself to bananas or anything that took his fancy. One night in 1956 a visitor awoke to find the Lord Mayor drinking water from a basin in his tent. Finding the water soapy, the elephant showed his displeasure by overturning the basin with a clatter. On another occasion he reached into a tent and absconded with underwear with which he dusted himself down.

A few nights afterwards this sagacious beast turned the tap of a stand-pipe which supplied the camp site, and had a drink. On leaving he failed to turn the tap off. To prevent a recurrence on the following night the water was turned off at source. When the Lord Mayor came for his drink and found the tap dry, he started in angry mood to tear up the stand-pipe. At this point the Warden, who was watching the proceedings, ran up and shouted, whereupon the elephant charged and chased him into a house. The following day an empty petrol drum was inverted over the tap, and packed round with stones at ground level. This arrangement baffled the Lord Mayor: he made a great deal of noise trying to get water; after a pause he lifted his trunk and struck the drum a resounding blow. He then stood quite still, as though deciding what to do next; then, with complete determination, he turned and made for the river. Unfortunately after various escapades involving visitors and their vehicles the Lord Mayor had to be shot.

Various characteristics

The Elephant is of course distinguished from all other African animals by a number of peculiar features: the vast body supported on pillar-like legs; the mobile, sail-like ears which when fanned to-and-fro act as radiators to dissipate heat; the tusks; and that extraordinarily versatile appendage, the trunk.

The trunk is a hand with sensitive prehensile fingers; and a strong arm to seize food and convey it to the mouth. It is used as a hose for a shower-bath; or – when looped to form a U-tube – as a drinking vessel. It is used to try the wind, to steer a calf, or to caress a mate. Variously it becomes a dust-spray, an inhaler, a weapon, a sand shovel in seeking water, a scanner of the olfactory news, and a musical instrument. All this makes the animal's plight more terrible when, as happens too often, the trunk is gripped in the poacher's wire noose and its distal part amputated.

The grey giants are by no means always grey. They take on the colour of the terrain where they have been mud-bathing. Spectacular –

The dust bath

especially when lit by the setting sun – are the red elephants of Tsavo West. In other areas they are brown, ochre, or a light chalky hue like that of the termite hills which they use as rubbing posts. But most, impressive, when they come clean from the river: then, while still wet, they are jet black and have a sinister look.

The normal gait is a free, swinging walk. Deceptively slow, owing to the length of stride, the walking elephant covers the ground at jog-trotting pace for a man. When a herd is seen moving through long grass, the backs heave and pitch like broad-beamed boats. The astonishing thing is that an elephant can move so silently; even in thick cover he disappears on cushion-like feet, leaving you to guess his whereabouts. When he is undisturbed and unhurried, there is gentleness and poise in every movement. An elephant passing through your camp at night will

King of beasts

even step over the tent ropes, leaving basin-sized footprints to show how he went by, unheard.

But at times elephants are the noisiest of animals. When frightened and stampeding the noise they make is appalling, as they crash heedless through obstacles with the force of a bulldozer. Their trumpeting and screamings then defy description: Temple-Perkins speaks of elephants 'trumpeting with that blood-curdling viciousness which, I have since learned, never fails to terrify when sounded at close quarters'.

Elephants are at home in any type of country – forest or bush, swamp or rocky hills. They can climb steep places with assurance. On occasion, as I have seen in the Kabalega gorge, they will descend seated and sliding on their backside. If pushed or angry, they adopt a fast shambling gait. The elephant's quickest pace – when fleeing or charging – is a trot: the legs, moving as fast as those of a trotting horse, then carry the animal forward on strides nine feet in length. Frederick Selous, the famous big game hunter of Central Africa, believed that an elephant, when going at its utmost speed, could cover 120 yards in ten seconds.

It is often said that elephants are unable to swim and that they cross stretches of water walking on the bottom. Herds are known to cross the Kazinga Channel at places much too deep for calves and adolescents to go over on foot. I once saw, and photographed, a mother and calf swimming in the Victoria Nile. Both were very low in the water, with the back almost awash and sometimes submerged, the ears folded back, and the tip of the trunk held erect, like a ship's ventilator.

It is a curious fact that oxpeckers, or tick birds – familiar associates of most of the large ungulates – are not found searching the hide of elephants for the ticks they harbour. For some reason elephants will not tolerate their attentions and the birds are of course readily dislodged with a swing of the trunk. But in Uganda a long-tailed, starling-like bird known as the Piapiac·(*Ptilostomus afer*) does associate with elephants and is commonly seen mounted on the head and back – sometimes a dozen or more birds taking a ride on the look-out for insects disturbed by their mount's progress through the grass. Another attendant wher-ever elephants are found is the Buff-backed Heron: one's attention is often first drawn to a distant herd by the sight of the conspicuous white bird riding high above the grass where elephants are feeding.

The mature bull elephant may weigh as much as six tons; and the female about four tons. In the field the sexes can be distinguished by the shape of the forehead and by the tusk-proportions. The male has a domed forehead; in the female it is angular in profile. Tusks of the male are broader at the base, proportionately heavier, and they taper from the root; those of the female tend to be narrow-based and parallel-sided. These differences are clearly seen in Plate 33.

Breeding maturity is reached at about the age of thirteen years; and the prime of life is between forty and fifty years. The gestation period is 660 days (22 months). Only one calf is born at a time. At birth the baby weighs about 265 lb and its shoulder height is under three feet: for nearly a year it is small enough to walk under its mother's body. The calf is weaned after two years but remains with the mother, so that a cow is often seen with two or even three calves of different ages in train. The interval between calving is not less than four years.

Elephant damage

Elephants require a great quantity of food. A large bull is said to eat about 900 lb daily. Field studies in which individuals have been under continuous observation by day and night have shown that about eighteen hours of the day are occupied by feeding. Elephants are very adaptable in diet and feed on a wide range of vegetable matter – grass, foliage, twigs, shoots, bark, seed-pods, fruit and roots. They will wade into deep water when feeding on papyrus, and are extraordinarily destructive – using their tusks as levers to break roots and fell timber, and killing large trees by ring-barking.

Destruction has been very severe in Tsavo National Park. This Park is situated astride the routes along which in former days the herds migrated between the Tana River and southern Tanzania. It is now the only area affording them sanctuary. The elephant population today greatly exceeds the carrying capacity of the habitat. Even in 1962 Noel Simon reported extensive damage over hundreds of square miles of country along the Voi, Tsavo and Athi rivers and elsewhere. 'Trees have been uprooted over a wide area giving the impression that a division of tanks has been manoeuvring there.' The destruction of Baobab trees is today widespread: the elephants rip off the bark, tear open the fibrous wood from ground level to a height of twelve feet, and fell great trees thirty feet in circumference.

In Kabalega Falls National Park elephants have laid waste great tracts of woodland. In some areas 98 per cent of the *Terminalia* trees are ring-barked and dead, and the habitat has changed progressively from diversified forest and woodland to open grassland. In an attempt to check this undesirable trend it was decided in 1965 to remove two thousand elephants by cropping.

This programme made available material for detailed studies of the ecology and reproduction of the elephant. One interesting aspect of this work, carried out by Dr R. M. Laws and other zoologists, was the discovery of a relation between high population-density and fertility. It was found that under population pressure, density-dependent controls of three kinds began to take effect. Firstly, the beginning of the reproductive cycle was retarded. Thus, while the age of puberty in the male was 11·5 years in a healthy population, in the population north of the Victoria Nile it was 15 years, and on the south bank – where pressure was greatest – breeding did not begin until 19·5 years. Secondly, there was found to be an increase in the interval between calving, from 4·5 years in a healthy population to 7·0 and 8·5 years, respectively. Thirdly, there was evidence of increased calf mortality (46).

In a slow breeding animal like the elephant, such built-in methods of population control – by lowered fertility, delayed maturity, and calf mortality – are slow to take effect. Meanwhile the damage continues and, as in Tsavo East, the elephants seem to be destroying the habitat on which their livelihood depends.

Intelligence and memory

The claim has been made that among comparable members of the animal kingdom learning ability is related to brain size: the bigger the brain, the greater the brain power. Such a generalization finds some support in the astonishing intelligence displayed by cetaceans – whales and dolphins – which is only just becoming clear from recent world-wide experimental studies of the animals in, and outside, marine aquaria.

The experiments suggest that cetaceans 'are second in intelligence only to man', and as Professor Teizo Ogawa of the University of Tokio says, 'In the world of mammals there are two mountain peaks. One is Mount Homo Sapiens, and the other Mount Cetacea'. If brain size is a rough indicator of mental capacity, one might expect to find a Mount Proboscidea somewhere on the map: elephants have the most massive brain of any land animal – more than thirteen pounds in weight.

The African elephant's mental attributes have yet to be scientifically studied. But owing to its long association with man as a working animal, a zoo inmate, and a circus performer, more is known of the Indian *Elephas maximus*. Fully trained working elephants in Mysore learn and obey some twenty-four different commands in Urdu, the equivalent of 'go forward', 'stop', 'lie down on your side', 'give me the object', and so on. Experienced animals are said to do their work with a minimum of commands, 'as if they "knew" what they were expected to accomplish'.

In 1957 Professor Bernhard Rensch (65) conducted a series of psychological experiments with a female Indian elephant in Münster Zoo, designed to test more precisely the animal's learning capacity and memory. Her initial task was to discriminate between various visual patterns. During training the patterns were presented in pairs, and the elephant had to choose the 'correct' one of each pair to get a food reward. The animal eventually learned and could keep in memory the meaning of twenty stimulus pairs. In a final discrimination test, each of the twenty pairs was presented thirty times in a previously established sequence. 'The test covered six hundred trials lasting several hours, yet the elephant not only showed no symptoms of fatigue but actually improved in performance towards the end.'

She also recognized patterns if they were modified. For example, having been trained to choose a black cross as the positive signal, she still recognized the figure as the positive sign when its shape was altered in various ways – suggesting her ability to grasp an abstract idea. A further series of tests in which she was presented with thirteen pairs of cards which she had learned earlier but had not seen for about a year showed, in a total of 520 trials, that she had retained the meaning of 24 (of the 26) visual patterns. 'This,' as Rensch remarks, 'was a truly impressive scientific demonstration of the adage that "elephants never forget" '.

Lt Col J. H. Williams, who worked with elephants in Burma for twenty-five years, reports that domesticated elephants have been known to stuff mud into the bells round their necks before silently setting out to steal bananas at night! And in his book *Elephant Bill* he says that the elephant 'never stops learning because he is always thinking'.

Black and White

To see for the first time a rhinoceros in the unspoilt and uninhabited wilderness is to be transported to another age, as if one were viewing the world in the middle Pleistocene long before man had arrived upon the stage. Its heavy build, median nasal horns, scale-like naked skin, small eye and dull wits all seem to mark it as coming from antiquated stock. Existing rhinoceroses are but a remnant of the number known from fossil remains. It is a family on the wane: in the world today there remain but five dwindling species, of which three are Asiatic and two African.

The Black (*Diceros bicornis*) and White Rhinoceros (*Ceratotherium simum*)

Portrait of Black Rhinoceros

differ from one another in many respects – anatomical and behavioural. Though the Black Rhinoceros is the smaller of the two, five feet at the shoulder as compared with at least six feet, it is nevertheless a massive animal that has fancifully been called a 'dreadnaught of the bush'. The prehensile upper lip is pointed and extensile, and can be used like a very short trunk to place food in the mouth – an adaptation to the browsing habit. The head is relatively short and is held horizontally, the neck is without a nuchal hump, and the ears are small and fringed thickly with hair.

The White Rhinoceros is distinguished by its much larger size with a weight of up to five tons, a pronounced hump on the neck, very long head which is carried low, large almost tubular ears, and broad square muzzle which – like that of the Hippopotamus – is well suited to the grazing habit. The horns of both species are variable in shape and length, but are generally longer in the Black Rhino: the record, shot by R. Gordon Cumming in South Africa, was a front horn measuring $62\frac{1}{2}$ inches.

The two species are very unlike in temperament and habits. The Black Rhino lives in bush country, especially thorn scrub where it eats coarse and prickly vegetation such as Gall Acacia or Whistling Thorn (*Acacia drepanolobium*). It will even take the fibrous woody fruit of the Sausage Tree (*Kigelia aethiopica*).

Black Rhino have been given a bad name for belligerence and vindictiveness by some hunters. No doubt this reputation is partly due to the animal's poor eyesight and inquisitive nature. When 'charging' the rhinoceros may be more alarmed than aggressive, and making an effort to flee from danger. Real aggressive intent is revealed by jerky movements of the head and tail, the animal standing with ears cocked and nostrils dilated. The trot is lumbering, but as the animal gathers speed its step has a spring, and it can stop and turn as quickly as a polo pony. The top speed is a gallop with the head down and the front horn held near the ground and, as Stevenson-Hamilton remarks, 'its appearance when bearing down to the accompaniment of snorts, reminiscent of a steam engine, is not reassuring' (74).

Black Rhinos tend to be unsociable. The males are generally seen alone. The female tends to lead a solitary life with the calf; and will not tolerate an older calf when the new offspring has arrived. If the calf strays the mother emits a high-pitched 'mew' – an incongruous call to come from so large an animal – which brings the calf to heel.

John Goddard, who made a three-year ecological study of two populations in Ngorongoro Caldera and the vicinity of Olduvai Gorge, found that adults remain permanently attached to a small home range of from six to fourteen square miles. The resident invariably attacks an intruder: 'The head is lowered, eyes rolled, ears flattened, tail raised, and the animal curls its upper lip, emitting a screaming groan. The anterior horns are used for goring, or for clubbing the other animal on the sides of the head' (30). Territory appears to be marked out by the use of regular defecating places. The sense of scent is extremely accurate: Goddard has shown experimentally that a rhino can accurately follow a devious fecal scent-trail; and one animal has been observed following another when separated by at least a mile, and going on an identical course.

Next to the elephant, the White Rhinoceros is the largest terrestrial animal. Despite its popular name, it is not lighter in colour than its

White Rhinoceros trotting

smaller relative: both species take on the colour of the mud where they have been wallowing, and so may be any shade of greyish-brown, chalky-grey, blackish, or red-ochre.

In temperament it is inoffensive and placid, almost always seeking safety in flight. A curious trait, in which it again differs from the Black Rhinoceros, is the mother's habit of walking behind her calf and guiding it with her horn; in the 'black' species the mother leads and the calf follows. The White Rhinoceros is also more sociable and is found in parties of three to five or more. Its distribution is discontinuous, in East and in South Africa, being restricted to a small area in the West Nile Province of Uganda and adjacent regions, and to game reserves of Zululand. Under protection in Umfolozi Game Reserve the species has flourished and since 1960 many specimens have been successfully immobilized, captured and transported to various reserves and zoos overseas. Specimens have likewise been translocated from the West Nile to Kabalega Falls National Park where they now live on good terms with the indigenous Black Rhino population.

White Rhinoceros, Umfolozi, Zululand

Rhinoceros horn is a structure unique among mammals. Unlike the horns of antelopes which have a central core of bone, or those of deer which consist entirely of bone and are deciduous, rhino horn is entirely keratinous, and situated on but not fused with the nasal bones. The horn is greatly prized by the Chinese and other eastern races for its medico-magical properties and especially for its supposed value as an aphrodisiac. It is also carved into various articles such as sword-scabbards, and libation cups which are believed to protect the user from poison. The high market price paid for the horn has long been a powerful incentive to poaching and illegal commerce – with Singapore as a main collecting centre for horns both from Africa and south-east Asia.

Published exploits of the Victorian hunters in Africa indicate the former abundance of rhinos. For example, when hunting in what is now Tsavo National Park, Captain Sir John Willoughby in 1887 'regularly shot four or five rhinos a day for many days on end close to the Mzima Springs. Forty-three rhinos were shot by his party within a fortnight' (71). The years have taken their toll; and today we are left with a decimated and dwindling remnant of these extraordinary animals.

The Hippopotamus

The last of the giants to be considered here is the Hippopotamus. The family to which it belongs is today confined to Africa, and is related to the pig family – Suidae. The members of both groups carry tusks, and have four toes on each foot. In the hippo the four toes touch the ground; in the pigs only the median pair do so. The Hippopotamus is really a gigantic pig that has become adapted to the amphibious life.

It is a grotesque animal, with huge head, enormously wide muzzle, bulging eyes, mouse-like ears, a barrel-shaped body and disproportionately short legs. The adult male may weight as much as four tons, or about the weight of an adult cow elephant. In profile the head has some features in common with the crocodile, and which are related to the amphibious habit: the nostrils are turret-mounted on top of the snout and can be closed with valves at the moment of submergence; the periscope eye is dorsal in position; and the ear opening is also situated high on the head – so that the animal can breath and remain alert and receptive when almost submerged.

Aggressive temperament

Hippos live in schools averaging ten individuals, but as many as 107 in a school have been recorded. The school contains adult females, calves and juveniles of both sexes, but few adult males. The latter tend to live

Hippopotamus in aggressive mood

apart and occupy wallows (45). By day the animals bath and doze in shallow water, grouped close together, one animal sometimes resting its chin on another's back. But their look of placid indolence is most misleading.

High specific gravity enables them to walk on the bottom when the body is quite submerged. They travel with great power and surprising speed when pursuing a rival or attacking a canoe or boat – not swimming but running under water, their progress made known by a displacement wave at the surface. During a fight which I once witnessed at Jinja the loser tried to make his escape under water, and for more than half a mile he was chased parallel with the shore, pursuer and pursued both totally submerged and only surfacing for air momentarily before resuming the underwater chase.

The males are aggressive fighters. Inter-male threat is demonstrated by the yawning display of the mouth and its fearful dental armament. At times the animals engage in tournaments, each charging and attempting to slash the flanks of its opponent. Battles – which take place at night – have been known to continue for many hours, the animals inflicting hideous wounds with the canine tusks, until one of the pair is mortally hurt. The hides of most old bulls are branded with scars. One bull which lived near Paraa had a raw wound on its side the size of a dinner plate. The astonishing thing is that such injuries do heal in time.

When disturbed on land the immediate reaction of a hippo is to make for its natural element, just as crocodiles do in like circumstances. If a hippo is encountered ashore by day, and you are caught on foot in its line of retreat, it is most likely to charge. The rule is: be vigilant and always pass them to landward. Hippos can travel very fast on their short legs, the gait being a quick trot rather like that of a warthog. Plate 31 well shows how one of these animals can cover the ground when pressed.

When observing crocodiles from a hide, unarmed, I have on two occasions been confronted by an aggressive hippo. Its demeanour is forbidding as it stands regarding one with a baleful eye, shaking its head in quick jerks and champing its jaws in a paroxysm of rage. The fact is, one soon learns to treat the Hippopotamus with due respect. When you are in camp these nocturnal visitors are not welcome. They have insatiable curiosity, and the camp fire attracts them. Like the Elephant, they can approach silently and stealthily, to make their presence known by a deep, loud bellow – 'Hoosh!', followed by a succession of shorter grunts – 'haw, haw, haw, haw.' This mirthless laughter is not pleasant to hear at close quarters.

Armed neutrality

Unless injured or old, the adult Hippopotamus has nothing to fear from any predator. The animal with which it is most likely to come into conflict is the Nile Crocodile. The two often meet and inter-mingle, both on land and in the water, and their relations are those of armed neutrality. On land the Hippopotamus is the acknowledged master: he has absolute right of way, and the reptiles readily give ground when one approaches, sometimes hissing remonstrance as they retreat.

At Fajao below Kabalega Falls I have watched hippos walking unconcernedly on the sandbar which is a favourite crocodile basking ground. If a crocodile failed to retreat the strolling mammal would deliver a blow with its muzzle, brushing the crocodile aside or knocking it into the water. Such an incident is illustrated in the drawing on page 145. At the same place in 1952 two males which were about to fight were seen

Mother and child

first to clear the arena by pushing all the crocodiles into the river. Stevenson-Hamilton states that at calving time either the mother or others of the school will drive all the crocodiles out of the pool in which they happen to be lying.

When protecting her calf in the water a female will snap at any crocodile that ventures too close. The young calf always keeps by its mother's side, and in crocodile-infested waters it has need to do so. The mother will sometimes carry its baby when in the water. In April 1972 when I was making frequent visits to a crocodile rookery, a hippo and calf were often seen bathing at a sheltered spot in lee of an islet, the young one squatting on its mother's broad back.

Sometimes a crocodile is overbold, and it may then be seized, and

even bitten in half, by the parent. In July 1956, near Paraa, I saw a crocodile lying in shallow water in two pieces – its body freshly severed in
front of the hind limbs. Encounters between hippo and crocodile, with
similar results, have been reported from Zambia and South Africa. No
one who has considered the span of a hippo's gaping jaws and its formidable fighting teeth will doubt its ability to inflict such damage.

A key species

Where present in large numbers, as in Uganda, the Hippopotamus plays
a major role in the ecology of inland waters. As a drainage engineer it
keeps down luxuriant vegetation bordering swamps, and maintains
water channels through papyrus and sudd. More important is the result
of their unique feeding behaviour. The feeding time is at night and in
the evening they go ashore along well-worn paths to graze – often
travelling long distances to fill the stomach, which holds about 450 lb of
grass. In the morning they return and discharge tons of manure into
the water. In this way they fertilize the lakes and water channels,
encourage growth of plankton and invertebrate life, and provide an
environment which can sustain fish and water birds in extraordinary
abundance.

In the fifties it became obvious that the Hippopotamus populations
of Rwenzori National Park had increased at a rate detrimental to the
ecosystem as a whole. Large areas of country near the shores of Lakes
Amin and George and the connecting Kazinga Channel were being
denuded of grass and eroded. In their nocturnal excursions hippos
were forced to extend their range farther from permanent water; and
many animals moved inland to occupy temporary rain-filled pans and
wallows, where they remained by day. Evaporation in dry weather
reduced smaller wallows to a stinking morass of mud where the crowded
animals floundered in attempts to keep their skin moist. Such a scene is
illustrated in the figure on page 96. At one of these drying pans I saw a
hippo that had been trampled right into the mud by its companions.

By 1960 the hippo population exceeded 12,000, and the Park authori-
ties decided to reduce the number by a carefully controlled shooting
programme. This provided invaluable research material; and as a result
of investigations by a team of zoologists and physiologists working at the
Nuffield Unit of Tropical Animal Ecology, led by Dr R. M. Laws, the
Hippopotamus became one of the most intensively studied mammals
in Africa.

One of the papers subsequently published was an account by Laws

The mud wallow

and Clough on the reproductive biology, based on post-mortem examination of two thousand animals. The life span was found to be about forty-three years. Males become sexually mature when seven or eight years old, females about a year later. Mating takes place in the water, and the young are born on land after a gestation period of 240 days. One young is born at a time: only two sets of twins were found in 276 pregnancies. At birth the baby measures about four feet in length and weighs a hundred pounds.

Births may occur at any time of year; but observations showed that the pattern of births is correlated with rainfall. Most conceptions take place in February and in August – namely, towards the end of dry seasons; and most young are born in October and in April – months of high rainfall, when the grass is fresh and rich in protein. Since the mother suckles her calf for twelve months, most calves will be weaned during the rains, when conditions are good for grazing (45).

5
Birds Great and Small

Nay more, the very birds of the air, those that be not hawks, are
both so many and so useful and pleasant to mankind, that I must
not let them pass without some observations.   *Izaak Walton*

T H E East African scene is renowned for birds in variety and for birds as
a spectacle: a mecca for the ornithologist or dedicated photographer;
and for the visitor whose interests are not specialized, something to
marvel at and long remember. In the National Parks of Uganda alone –
Kabalega Falls straddling the Victoria Nile, Rwenzori in the south-west,
and Kidepo Valley in the remote north-east – a total area of some 2,800
square miles (or roughly the size of Devonshire) there have been recorded over seven hundred species of birds – some two hundred more
than on the British check-list.

Families endemic to the Ethiopian Region

Certain endemic African birds are of special interest in that they belong
to families which do not occur outside the Ethiopian Region. One of
these is the Hammerkop (Scopidae), a sombre brown bird that is always
found near water, where frogs are its favourite food. The nest, built in
large trees, is an enormous fortified structure, three or four feet high
and broad, made of thorn sticks cemented with mud and grass. An
entrance at the side leads by a tunnel to the mud-lined brood chamber.
Natives fear the bird, believing it to have evil powers.

The Whale-headed Stork (Balaenicipitidae) is a tall, grey bird with a
massive dilated bill which distinguishes it from all others. Its haunts are
the sudd in southern-Sudan and the swampy borders of lakes in Uganda,
where it is known to feed on lungfish.

Like the preceding two species, the Secretary Bird (Sagittariidae) is
the only representative of its family. On casual acquaintance it has been
described as 'a large bird with a head like an eagle and very long legs
like a stork'. Standing over three feet in height, it has a wing-span of
seven feet. The dark feathers which protrude from the back of the head,
the strong raptorial bill and the long tail feathers give the bird a distin-

guished look as it walks about with measured strides in search of the reptiles which are its main food.

The Wood-hoopoes (Phoeniculidae) and Mousebirds (Coliidae) are again two families whose members are confined to the Ethiopian Region. The former are scimitar-billed, long-tailed birds, with dark lustrous plumage. The sparrow-sized Mousebirds have a very long tail, short rounded wings, and are dull brown in colour; in flight they are remotely suggestive of miniature hen pheasants. After alighting they scramble along twigs of bushes using the bill as well as the feet – their progress sometimes appearing more mouselike than avian.

In another endemic family are the Turacos (Musophagidae). Principally found in evergreen riparian forest, they are fruit-eating birds, distinguished by their fine colours and by their raucous cries. Guinea-fowls (Numididae) – familiar from the now-ubiquitous domesticated bird – again belong to a group exclusively African. The most beautiful member of the family is the Vulturine Guineafowl (*Acryllium vulturinum*), which is thus described by Bannerman: 'The bare skin of the head and neck is greyish-blue, a horseshoe band of downy chestnut feathers ornaments the nape, the breast and abdomen are cobalt-blue; long pointed hackles, black with white shafts and margined with cobalt, fall from the lower neck and mantle, while the rest of the plumage is black, spotted with white. Very characteristic is the tail of sixteen feathers, the middle ones considerably elongated and pointed. The crimson eye lends a wicked look to this truly magnificent member of the family.' It lives in arid bush; and bands of the birds are always to be seen near Samburu.

The Ostrich (*Struthio camelus*), sole member of the family Struthionidae, is distinguished from all other birds by possessing only two toes, and it is of course the largest living bird – the adult male weighing about 300 lbs and standing nearly eight feet in height. It is a roving omnivore, an inhabitant of open plains and bush. It provides an extreme example of cursorial adaptation among birds: its unfeathered legs and thighs are long and exceptionally powerful, and a frightened bird has been seen to overtake and shoot ahead of a hartebeest going at full gallop. A bird so keen-sighted and swift of foot has little to fear from any predator; but ostriches are sometimes killed at the nest by lions, and the eggs are eaten by hyaenas, jackals and the Egyptian Vulture – the latter throwing stones to break the thick shells.

Birds and bird names

For some species the names of gems or precious stones have been used to suggest brilliant colours and fiery iridescence of plumage – Emerald Cuckoo, Malachite Kingfisher, Amethyst Sunbird, Pearl-spotted Owlet. The Emerald Cuckoo is bright metallic green, with a yellow belly. John Williams, in his invaluable field guide (85), calls it one of the most brilliantly coloured birds in Africa. The Malachite Kingfisher, always to be seen at close range from a launch in the Kazinga Channel, on the Victoria Nile and on Lake Naivasha, as it perches on a reed or overhanging bough, has a cobalt, black-barred crest, back and wings of ultramarine blue, chestnut underparts and bright red beak and feet – a strikingly beautiful bird at rest or in flight. The sunbird is one of many Nectariniidae which, in the male, specialize in scintillating and fiery colours; the Amethyst is one of the dark species, velvety-black with a green cap and purple throat. The owl is of course in a different category; like its congeners it is sombre-hued, its 'pearls' merely white spots on the soft feathers of wings and tail. This is a charming mini-owl, standing only some five inches high from its perch. It is often seen about in the daytime, when it attracts the mobbing reactions of other birds, and so unwillingly draws attention to itself.

Many birds have been given names that hint at some aspect of their life: such are Palm Swift, Ant-eater Chat, Flappet Lark. Wherever there are Oil-palms, Dom-palms and particularly the massive Borassus (*B. flabellifer*) one looks for the Palm Swift, for there is a definite association between bird and tree. Before nesting time parties of swifts circle and scream round the crown of leaves, where later the tiny nest – a shallow cup made of feathers – is glued with sticky saliva to the under side of a lower drooping palm-frond. The chat *Myrmecocichla aethiops*, as its scientific name implies, is myrmecophagous, though in fact the bird subsists mainly on termites rather than ants. A large black chat with white wing-patches conspicuous in flight, it is common in the Eastern Rift Valley and always to be seen in Ngorongoro Crater. The Flappet Lark is a bird to surprise the uninitiated, for it uses instrumental effects as an auxilliary to song. In the nesting season the male produces burst upon burst of applause, clapping its wings together as it flutters high over its territory – with a sound of far distant machine-gun fire.

Of course other trivial names with ecological connotation are usefully applied to whole families or groups of species – babblers, bee-eaters, flycatchers, weavers, kingfishers and so on. Members of the first group associate in noisy parties, giving voice to babbling and chattering notes. The Black-lored Babbler is one often seen at bird-tables in the garden of Lake Naivasha Hotel. Weavers are well-known for their skill in making woven, domed, basket nests whose architecture varies from species to species. Some, like those of Buffalo Weaver, are large untidy structures, generally found in the branches of Baobab. Others, as in the Grosbeak Weaver, are neat, compact globes made of the green threads taken from Papyrus crowns. Some again are globular cradles of grass, with a lower entrance porch separated internally from the egg chamber. Others are furnished with very long hanging entrance-spouts – which may serve as an anti-snake device. Weavers often make their nests at the extreme tip of palm-fronds or at the end of horizontal branches of acacias, and so deny foothold to birds'-nesting vervet monkeys and baboons. The bee-eaters and flycatchers are of course insectivorous; but kingfishers, despite their title, are not necessarily piscivorous – in fact, most species do not fish. Many thrive far from water and eat insects. The Grey-headed Kingfisher, for example, can be seen in dry semi-desert bush of Karamoja in northern Uganda.

To those less familiar with the wonderful East African avifauna, the names of some birds will arouse speculation. What, for instance, are we to make of Crombec or Piapiac, Purple Grenadier or Tropical Boubou, Cardinal Woodpecker, Zanzibar Red Bishop, or Cut-throat. The bishop, to be sure, is in good company with other dignitaries – for there is a Black Bishop, a Yellow Bishop and a Fire-fronted Bishop, to mention but a few of these lovely weavers whose males wear bright vestments. The Cut-throat is a small seed-eating bird whose sinister name derives from the crimson throat-band worn by the male. And there is nothing martial about the Purple Grenadier, a waxbill related to the Red-cheeked Cordon-bleu; the two may often be seen feeding together in short grass.

Many names serve to remind one of the explorers, administrators or naturalists who introduced the species to science – Abdim's Stork, Ruppell's Griffon Vulture, Verreaux's Eagle, Livingstone's Turaco, Fischer's Lovebird, D'Arnaud's Barbet, Von der Decken's Hornbill, Hildebrandt's Starling, Speke's Weaver, Jackson's Widow-bird. Again, the names of other species seem to have a magic of their own, and act as magnets to draw the visitor to their haunts – Paradise Flycatcher and Sacred Ibis, Bronze Mannikin and Silverbird, Augur Buzzard and Pale Chanting Goshawk and Crowned Crane are just a few that come to mind.

Perhaps surprisingly, there is poetry even in the scientific names

Crowned Crane

bestowed on species by the taxonomist. *Lamprocolius splendidus, Spreo superba* and *Cosmopsarus regius* are three magnificent starlings. The first is a bird of the forest canopy, often to be seen in flocks in the Botanical Garden at Entebbe. With plumage of brilliant metallic-green and blue, this 'splendid, radiant jackdaw' is aptly named. The other two, the Superb and Golden-breasted Starling, will be very familiar to travellers who have stayed at Kilaguni in Tasvo National Park, where both species, along with Buffalo Weavers and hornbills, have become so 'tame' as to enjoy taking after-noon tea with the tourists. The Golden-breasted Starling, in build somewhat like a miniature pheasant, has metallic blue-green upper parts and a belly of gold. Williams says of this bird: 'The most beautiful of the East African starlings' – a compliment indeed! Such a creature, under continuous and close observation, stirs in one a sense of the marvellous. We look in wonder – can one really believe one's eyes? – a philosophical question more easily asked than answered. When in sight of such things

I am reminded of a remark I once overheard in the aquarium of the London Zoo. A mother and little girl were standing by a sea-water tank in which were cruising an assortment of brilliant reef-fishes. The question was – 'Oh Mummy! Are they real?'

Another name I have always liked is *Limnocorax flavirostra*. The trim little moorhen to which it belongs – black of plumage, with red iris, red legs and a lemon-coloured beak – lurks among reeds or papyrus at the waterside and may often be seen feeding among water lilies by river-bank and lake-shore; how appropriate its name, 'the marsh crow with the yellow bill' – familiarly called the Black Crake.

A ball of birds

There is something compelling about the sight of birds congregated compactly – birds in a vast crowd. When in addition the individuals are themselves colourful and the flock extends right round the shore of a lake the size of Lake Nakuru, the spectacle is tremendous. Roger Tory Peterson, the renowned ornithologist and artist, described the Nakuru flamingos as 'the most staggering bird spectacle in my thirty-eight years of bird watching'. From careful sampling counts, the numbers present are believed at times to exceed a million birds, comprising perhaps a third of the entire Lesser Flamingo population of East Africa. There is a good deal of movement from lake to lake in the Rift Valley, so the number tends to fluctuate in any one place. Periodically Lakes Hannington and Elmenteita are great gathering places; and there are seasonal movements to Lake Natron where both Greater and Lesser Flamingo breed. Another favourite spot in Tanzania is the shallow lake in Ngorongoro Crater. Looking down 2000 feet from the escarpment rim, one can see in favoured spots round the shore-line a kind of pink bloom or powdering of colour, where the birds in thousands stand crowded in the shallows.

But sheer numbers, even without the added beauty of plumage, make an indelible impression on the mind. In the lower Victoria Nile migrating Abdim's Storks provide such a spectacle. In April great flocks of these birds appear when in transit from southern Africa to the northern savannas where they breed during May to September. After a morning feed on grasshoppers the birds congregate towards midday on mud-banks and sand-rivers in their multitudes. Having rested, they take off, to rise high in thermals, hundreds together circling in great spirals to gain altitude for a glide northwards that will take them another stage on their way.

Globe of Red-billed Quelea

But more remarkable still are the massed formation flights of the dull-hued, sparrow-sized Red-billed Quelea or 'Sudan Dioch' (*Quelea quelea*). My first introduction to this bird was in 1952 from a fisheries launch on Lake Victoria. We were cruising close off-shore where a belt of forest came to the water's edge, when suddenly from out of the canopy appeared a cloud-like swarm of little birds, flying rapidly over the trees in a compact body. It was as though, once they had assembled in formation, an invisible elastic envelope encompassed them from without, to exert some intangible pressure on the whole flock so as to compress the birds into the smallest space compatible with flight. Meanwhile, each individual maintained the necessary flying room and in some miraculous way the several birds seemed to be evenly distributed within the flock. Thus, while each kept apart from its neighbours in a sort of aerial territory, the group as a whole moved like a compact animated ball.

A remarkable feature of this exhibition was the speed of flight and precision of movement. The congregation, which comprised some thousands of individuals, rose and swung and turned as a unit. There appeared to be no leader: there were no stragglers in this rolling, surging, plunging sphere. As we watched, the birds continued on a course parallel with the shore, and then it was that the astonishing thing happened, for from out of three trees – giants standing high above the rest – there rose three more bird-clouds, each in turn joining and being absorbed by the original group as it passed over their perching place, so that by successive increments the speeding ball of wings merely increased in size without changing shape. The whole flock then rose high and out over the lake, the specks within the mass turning and twisting as water-logged particles might move in a glass globe. Sometimes the globe itself would be distorted as an oval, or would open and close like a concertina, but always to return to its proper spherical shape. On another occasion when we were perhaps half-a-mile off-shore the bird cloud lifted from the trees like smoke such as might be caused by a heavy explosion. The mob of birds moved over the lake. As the individuals wheeled and turned in perfect unison, the appearance of the flock changed and changed again – at one instant darkening as the innumerable silhouettes banked over to be seen in broadside view, at the next the birds would wheel away and momentarily vanish as if vaporized, to reappear again in clear relief. The hurrying, pulsating globe then veered off towards the trees, and could no more be seen, except for an instant when in a final evolution all the wings simultaneously caught the sunlight.

It is only in recent years that *Quelea quelea* has acquired its reputation as the most numerous and also the most destructive bird in the world – a kind of avian locust. In 1952 Kenya first became aware of the threat to wheat-growing localities in the Rift Valley Province; and John Williams, then Ornithologist of the Coryndon Museum, Nairobi, carried out a survey in the Turkana region. His report (84) published in 1954 contains valuable information on the biology of these prolific weavers. Their powers of reproduction are highly remarkable. The birds may breed at any time of year; they are polygamous, with a sex-ratio of four females to every male; and they become sexually mature before they are a year

21. Goliath Heron, Victoria Nile, Uganda

Nesting Colony of Sacred Ibis, Buligi, Uganda

22. Yellow-billed Storks, Lake Nakuru National Park

23. Buff-backed Heron, near Kabalega Falls

Secretary Bird, Lake Nakuru, Kenya

old. Williams calculated that under favourable climatic conditions which enable the birds to produce five broods in a year, 'a colony of 1000 birds would increase to over 14,000 in one year, even allowing for 10 per cent mortality'.

Queleas normally feed on wild grass seeds in remote thorn-bush country, where they breed. But if a succession of good breeding years is followed by drought the birds must migrate or perish, and it is then that they descend upon farmlands to attack cereal crops of rice, wheat and millet. The concentrated swarms affect the economies of countries across the continent from Senegal to Ethiopia and Somalia, and from the Sudan southwards through Kenya, Uganda, Tanzania and Zambia to southern Africa.

The size of the raiding swarms may be imagined when ecologists refer to ten thousand birds as a small flock. One of the largest congregations ever seen was in southern Tanzania where a roost, covering forty square miles, contained an estimated twenty-five million birds. A flight of two or three million queleas seems to be considered fairly average.

Many experiments have been conducted to discover suitable methods of control. Traps, poisonous aerial sprays, dynamite, petrol bombs and flame-throwers have all been used against the birds which roost at night in trees near the grain fields. It is recently reported that in Senegal two destruction teams killed eighty million queleas in five months; and in South Africa queleas have been destroyed at a rate of a hundred million a year. Yet the birds return in even larger numbers.

Hornbill habits

We may pause here to take a look at the hornbills, which on account of their remarkable nesting habits are of unique interest. The hornbills are an unmistakable group, characterized by the enormous bill which gives the bird a top-heavy appearance when on the wing. In many, and especially in the larger species, the upper mandible is surmounted by a casque. This structure is very pronounced in large forest-dwelling East African species – such as the Trumpeter (*Bycanistes bucinator*), Black-and-white Casqued (*B. subcylindricus*) and Silvery-cheeked Hornbill (*B. brevis*). But despite its clumsy appearance the casque is in fact a light structure of cellular bony tissue within a shell of horn.

Hornbills are of course not to be confused with those other enormous-beaked tropical birds – the toucans. Hornbills are members of an order (Coraciiformes) which includes the kingfishers, bee-eaters, rollers and hoopoes; and are an old-world family ranging from Africa to the

Solomon Islands. Toucans (Ramphastidae) are confined to the new-world tropics, with their headquarters in the Amazon basin, and are related to barbets, honeyguides and woodpeckers (Piciformes). The larger hornbills, and the toucans, are both frugivorous; and the two groups provide an interesting example of convergent evolution. We may notice, in passing, that in beak size some toucans outdo the hornbills, having a beak even exceeding the body in length.

In their breeding habits hornbills differ from all other birds. Many wonderful and inaccurate tales about them were brought back to Europe by early travellers: it was popularly believed that the males imprisoned their mates in the nest-hole; that they fed the gizzard lining to them; and that they eventually died from the effort of looking after the family. The facts – established for several of the East African species by the painstaking observations of Moreau (55, 56) – are different, but not less remarkable.

Silvery-cheeked Hornbill

The Silvery-cheeked Hornbill is a large forest species, black-and-white in colour, with a powerful voice like the bray of a donkey. It is strictly frugivorous, feeding on berries, wild figs, and small stone-fruits. In October the pair investigate possible nesting trees for some weeks before

Silvery-cheeked Hornbill

selecting a high nesting hole, which is usually one with the outer entrance narrower than the internal cavity. The male takes an interest in the nest site, peering into the hole but never entering it.

Then begins the work of plastering. Each morning the female enters the hole. The male collects a load of soil, flies to the entrance, and regurgitates, one by one from his gullet, pellets of clay mixed with copious saliva. These pellets are presented by and received in the bill. The female slaps each pellet on the entrance and works it with the bill, as with a mason's trowel. The material solidifies to form a tough barrier. For the plastering, there is complete division of labour – the male brings the soil and cement, and the female fashions it. Day by day the hole diminishes until the morning comes when the female enters with difficulty and after work is unable to leave. Without coercion by the male, she has walled herself in; and there begins for her a sixteen weeks' period of imprisonment.

Thereafter the plastering continues until only a narrow vertical slit remains – large enough to admit the tip of the male's bill when feeding his mate. In the hornbills the number of eggs in the clutch tends to vary inversely with the size of the bird; and in the present species the number rarely exceeds two, and often only one egg is laid. The female remains in the hole not only during incubation, but until the young are ready to fly. Of necessity the supply of food for mate and young devolves upon the male bird. The magnitude of this task is shown by Moreau's dawn-to-dusk observations in the Usambara mountains in eastern Tanzania. From nine to twenty-seven visits are made by the male to the nest each day, the number increasing as the young develop. At each visit he brings up to 45 berries in the crop: these are passed by regurgitation one by one to the female. During the whole nesting cycle a male may bring, in about 1600 visits, not fewer than 24,000 fruits. Not surprisingly, as Moreau remarks: 'The male's industry slackens in the last fortnight of 15 to 17 weeks' labor'. When the young are about eight weeks old the female breaks away the plaster and flies out, followed by the nestlings – all emerging on the same day, and able to fly well from the moment of leaving the hole.

Insectivorous hornbills

In nesting habits the smaller hornbills of the genus *Lophoceros*, which include the Red-billed (*L. erythrorhynchus*) and Von der Decken's Hornbill (*L. deckeni*), so familiar to visitors to Kilaguni Lodge in Tsavo National Park, differ in certain respects from the large forest species. In the first

Red-billed Hornbill

place they are insectivorous, and insects, unlike a cargo of berries, have to be captured and transported one at a time to the nest-hole. Moreover, the family is larger, the normal clutch numbering three to five eggs. These two factors would place an intolerable burden on the male were he obliged to feed his family until the young were fledged. The difficulty is met by the earlier emergence of the mother. Having completed a rapid moult, she breaks out of the nest when the young are still only a third grown; thereafter she assists the male in feeding them.

But most remarkable is the behaviour of the young at this time.

After the female has flown out the young birds, less than four weeks old, themselves set to work and elaborate plaster from rotten wood, dung, and insect remains bound with saliva. This material they place on the hole-side and hammer it with the flat side of the bill tip, until the damage to the entrance caused by the mother's departure has been repaired. At a later stage, after their flight feathers have grown, they themselves break out of confinement to join their parents.

The specialized habits of hornbills obviously call for special sanitation. Techniques which promote nest hygiene are of three kinds. High-velocity defecation – made possible by the well-developed cloacal muscles – is a habit of many hole-nesting birds. In the hornbills, the female climbs up the nest cavity to expel the faeces with surprising force through the slit-like entrance. Accumulations of dung extending several feet away from the base of the tree often provide the first clue to the whereabouts of a nest. Secondly, at intervals she deliberately gives the nest a 'spring clean', ejecting with her bill feathers, egg-shells, fruit-stones and bits of rotten wood. Occasionally the male has been seen to

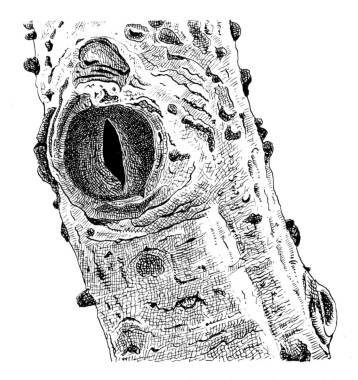

Nest hole of Yellow-billed Hornbill (based on a photograph by North)

help from outside, receiving the rubbish piece by piece through the slit, and dumping it. Thirdly, sanitation is assisted by a considerable fauna of scavengers resident in the nest cavity, including various scavenging beetles, moth larvae, bugs and omnivorous wingless cockroaches. The clean condition of the nest right up to the time the young are able to leave must owe much to the activities of these commensals.

Like the nesting arrangements, the female hornbill's method of moulting is also unique among birds. Correlated with the period of close confinement, which precludes flight, the mother undergoes a curiously sudden moult. In some species this takes place so quickly that within a few days of settling in the hole she is almost without flight feathers and indeed practically naked. In *Lophoceros* the moult is catastrophic and begins soon after the first egg has been laid. New plumage sprouts soon after the young have hatched. The female is completely flightless for a period of some eight weeks, and of course her dependence upon the male at this time is absolute.

The Ground Hornbill

The Ground Hornbill (*Bucorvus leadbeateri*) is a very large bird, appearing black, and only showing its white primary feathers in flight. Its colour is enlivened by the bare skin round the eye, and on the chin and throat, which in the male is bright scarlet. It is the most terrestrial of the hornbills, having the tarsus elongated, and the tail short. When feeding the bird struts about like a turkey, searching for fruit, insects, small mammals, young birds, frogs and reptiles. Tortoises are much relished; and with its bill the bird can neatly extract all the flesh from the unhappy reptile, including the head and legs – leaving the shell clean and undamaged. An observer in Zambia has described the hornbill's method of dealing with snakes: 'The head was gripped and crushed in the beak, and dropped; the mid-body was grasped in the beak, shaken violently, and dropped; the crushed head was grabbed and the body thoroughly crushed by being moved sideways through the beak from end to end several times; the prey was dropped and stabbed with the point of the beak at close intervals for the whole length; it was then swallowed head first.'

These hornbills lack the peculiar breeding habits which distinguish members of the Bucerotidae from all other birds. The site selected for nesting is a large hollow or depression in a tree such as that formed at the end of a broken limb. Owing to the nature of the nest no plastering is attempted, and the female – not subject to or protected by restricted quarters – remains active during the nesting period. The social structure of ground hornbills has recently been studied in the Kruger National Park by Kemp (37), who found that the foraging groups, numbering up to eight individuals, each include an adult female who dominates all other members, males and females alike, and is fed by them all when she is incubating.

In many parts of Africa the Ground Hornbill is an object of superstition. Kaffirs formerly held the bird sacred, believing it to possess magical properties. It was only killed in time of drought and then by order of the 'rain doctor'. The bird has an extremely offensive smell; and it was believed that when the corpse had been thrown into a river it would make the river sick, and heavy rains would come to wash it away.

The bird was also put to more practical use in the days before firearms were possessed by native hunters. A hornbill's skin, with the head and neck intact and wired up in life-like attitude, was placed over the hunter's head and shoulders; thus disguised, he would crawl towards his quarry through the grass, imitating the movements of a living ground hornbill until he was close enough to use his bow.

In East Africa there is an interesting association between the Carmine Bee-eater (*Merops nubicus*) and the Ground Hornbill, similar to that between the Cattle Egret and elephant or buffalo. The bee-eater rides on the hornbill's back, capturing on the wing insects that are disturbed by its mount's progress through the grass. It is interesting to note that this bee-eater likewise patronizes the Kori Bustard. Writing in 1898 Arthur Neumann described this as a common sight. 'It sits far back on the rump of its mount as a boy rides a donkey. The pauw does not appear to resent this liberty but stalks majestically along, whilst its brilliantly-clad little jockey keeps a look-out, sitting sideways, and now and then flies off after an insect it has espied, returning again after its chase to its "camel".'

6
Birds of Prey

Doth the eagle mount up at thy command, and make her nest on
high? She dwelleth on the rock, and hath her lodging there, upon
the crag of the rock, and the strong hold. From thence she spieth out
the prey; her eyes behold it afar off. Her young ones also suck up
blood: and where the slain are, there is she.   *Book of Job*

THE term Birds of Prey, used for this chapter heading, can have different
meanings. In terms of feeding habits, all birds must be included – other
than those that are graminivorous, frugivorous or herbivorous – be
their prey never so varied. On the other hand, to the ornithologist the
term is used in a taxonomic sense, and refers to members of the order
Falconiformes, which includes the vultures, and the true raptors –
falcons, kites, eagles, buzzards, hawks and harriers. The term can also
be used to embrace the nocturnal predators, or owls – members of the
order Strigiformes.

In an ecological context the distinction becomes blurred where we
find similar but unrelated birds leading a like mode of life. Such a case
of convergence in form and habit is seen, for example, in the shrike-like
Pygmy Falcon and the falcon-like White-crowned Shrike: both birds
are insectivorous and both may be found in the same habitat. Again,
particular prey species may be taken by birds of very different habits and
affinities. It can matter little to the individual *Tilapia* whether its slayer
be a Fish Eagle, a Goliath Heron, or a White-breasted Cormorant.

In the following pages we shall refer mainly to the feeding habits of
some of the larger East African birds, dealing first with what may broadly
be called 'water birds' – including the cormorants, darters and their
allies (Pelecaniformes), the storks and herons (Ciconiiformes) and
certain highly specialized feeders. In the second part of the chapter we
shall look briefly at the feeding techniques and structural adaptations
of the diurnal and nocturnal birds of prey.

Cormorants and Darters

It is almost axiomatic that two or more related species cannot occur in
the same habitat unless they achieve some form of ecological divergence,

African Darter or Snake Bird

which tends to reduce direct competition between them. This concept of ecological separation is illustrated by many striking examples among African birds and mammals. Thus, the Black and White Rhinoceros, and the Tree and Rock Hyrax, are two 'species-pairs' in each of which one member is a browser and the other a grazer. Greater and Lesser Flamingo are again differentiated by their feeding habits (see page 123).

Of special interest in this connexion are the cormorants and darters of fresh water lakes and rivers in East Africa. At first glance the feeding habits of the White-breasted (*Phalacrocorax lucidus*) and Long-tailed Cormorant (*P. africanus*) and the African Darter (*Anhinga rufa*) might appear to be generally similar. These foot-propelled diving birds are often found in the same waters; unlike the herons (which stand and wait) and the kingfishers (which plunge from above), they dive from the surface, and all three are exclusively fish-eaters.

I had good opportunities of watching and examining the species in Uganda during a year's field work in 1952. Stomach contents of birds collected from Lake Victoria, the Victoria Nile, Lake Mobutu and Lake George showed marked feeding segregation. As will be seen from the following list of prey, no fewer than twenty-six, out of thirty-one genera or species represented, are exclusive to the diet of one or other of the three predators.

Fish prey	White-breasted Cormorant	Long-tailed Cormorant	African Darter
Mormyrus kannume	●		
Mormyrus sp.	●		
Mormyrops anguilloides			●
Bagrus sp.			●
Synodontis sp.		●	
Alestes sp.		●	
Engraulicypris argenteus	●	●	
Engraulicypris bredoi		●	
Lates sp.		●	
Aplocheilichthys sp.		●	●
Aplocheilichthys pumilus			●
Tilapia sp.			●
Tilapia nilotica			●
Tilapia zillii		●	
Haplochromis michaeli	●		
Haplochromis gracilicauda	●		
Haplochromis guiarti	●		
Haplochromis 'erythrocephalus'	●	●	
Haplochromis perrieri		●	
Haplochromis mahagiensis		●	
Haplochromis wingatii		●	
Haplochromis albertianus		●	
Haplochromis obliquidens		●	●
Haplochromis nubilus		●	●
Haplochromis multicolor			●
Haplochromis squamipinnis			●
Haplochromis edwardii			●
Haplochromis nuchisquamulatus			●
Haplochromis schubotzi			●
Haplochromis pharyngomylus			●
Haplochromis serranus			●

Occurrence of fish prey in stomachs of 56 White-breasted Cormorant, 112 Long-tailed Cormorant and 68 African Darter. (Species represented by one occurrence only are omitted.)

These differences in prey are reflected in the divergent feeding techniques. *Phalacrocorax lucidus* tends to fish more in open water – often in Lake Victoria beyond sight of land. The birds have well-defined and extended flight-lines, used particularly at dawn and dusk, between roosting-rookeries and feeding-grounds, and in flight they are gregarious. The other two species feed closer in-shore, fly singly, generally within easy range of the roost, and without the marked diurnal flight-rhythm of *P. lucidus*.

When diving *P. lucidus* assists the deep dive by leaping clear of the water before plunging head first. *P. africanus* dives, grebe-like, from a swimming position. *Anhinga rufa*, whose habits are the most specialized, submerges the body slowly, like a diving submarine, the neck and head extended above the surface being the last part to disappear. Darters are heavy, and when dead they barely float and often sink. When swimming at the surface, the body is more or less submerged, and to this the 'snake-bird' owes its popular name.

In a sequence of fishing dives, the pattern of submergence and surfacing times differs in the three species. *P. lucidus* takes the shortest breathing time at the surface between dives – generally 3–5 seconds. *P. africanus* is a much smaller bird – often called the Pygmy Cormorant; and has the shortest diving times, which average 14 seconds. *A. rufa* tends to stay longest under water, with a mean submergence time of 30 seconds, and maximum timed dive of 92 seconds.

The cormorants pursue their prey under water and catch it in the strongly hooked bill. The darter's hunting method is extremely interesting. The bird approaches its prey, often quite slowly, with the head retracted until within striking distance. It then suddenly straightens the kinked neck and *impales* a fish on its slender lance-like bill. It is sometimes stated that the bird catches its prey between the mandibles. But examination of fish recovered from the crop leaves no doubt as to the method used. Small fish are sometimes speared with one mandible: larger prey show a double wound where the partly separated mandibles have transfixed the body from side to side. The disabled fish is then brought to the surface, and, with acrobatic skill, so thrown into the air that it falls to be caught and swallowed head first.

The freshwater pelicans, of which there are two species in East Africa – White (*Pelecanus onocrotalus*) and Pink-backed (*P. rufescens*) – do not dive, like the cormorants, but fish from the surface. A characteristic feature is the capacious bill; the upper mandible is long, narrow, flattish and hooked at the tip; the lower supports a widely distensible gular pouch, used as a landing-net when the bird is fishing.

Pelicans fish in well-organized parties, numbering from seven to a dozen individuals. In open water, as on Lake Nakuru, the birds swim slowly in line ahead or crescent formation, sometimes beating the water with their wings to drive the fish before them, and then, as though in response to some signal, all bills are plunged below the surface simultaneously. When fishing along a wall of papyrus, the formation will wheel round to trap the shoal in an animated seine net.

Herons and Storks

The Goliath Heron (*Ardea goliath*), much the largest of the African Ardeidae, is an uncommon, local resident on rivers, lakes and lagoons. Essentially a fish-eater, it is only found near water. When feeding the bird stands motionless in the shallows, waiting until its prey swims within striking distance of the powerful, spear-like bill. Being a very tall bird it can of course fish in deeper water than the Grey Heron. In

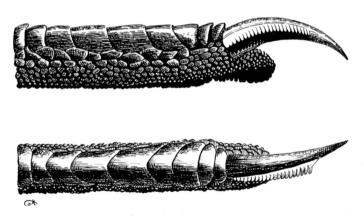

Digital comb of Goliath Heron

Uganda, *Polypterus*, *Barbus* and *Tilapia* are included in its diet. And like the Marabou and Saddle-bill Stork, it is known seasonally to take crocodile hatchlings.

A curious structural adaptation found in the herons is the pectinated claw of the middle toe. The Goliath has this claw fashioned into a sizeable comb, which I have illustrated in side and dorsal aspect on this page and which is used in feather maintenance. Owing to the nature of its food, the heron's head and neck tend to become fouled with slime:

this is rectified by appropriate use of cosmetics. The bird first rubs the soiled feathers on the 'powder-down' patches of the body where disintegration of feather barbules produces a dry bluish powder. Next the powder and the slime are scratched off with the toe-comb. Finally, to complete the toilet, the bird anoints and dresses the disarranged feathers with oil obtained from the preen gland at the base of the tail.

Incidentally we may note that various other slimy-feeders, such as cormorants, gannets and finfoots, have the claw of the third digit similarly modified; while in the pelicans, and in the snail-eating Open-bill Stork, instead of a comb the inner border of the third claw has become a sharp-edged scraper.

The Purple Heron (*A. purpurea*) frequents papyrus and reed beds bordering swamps – an inconspicuous bird as it adopts its cryptic pose, standing motionless, bolt-upright with the bill pointing to the sky. It feeds on fish and frogs; in Uganda the *Hyperolius* tree-frogs are a favourite prey.

The Squacco Heron (*Ardeola ralloides*) is generally found in lagoons and marshes where shallow water is covered with vegetation. In Lake Naivasha the floating leaves of water lilies are a favourite haunt, where the bird squats, huddled up and motionless, waiting for prey to approach. Specimens I examined in the south of Lake Mobutu and in the Damba-Kome Channel of Lake Victoria had been feeding on species of *Engrauli-cypris* and *Haplochromis*. Sir Frederick Jackson has described how at Entebbe it was seen feeding on a Cichlid fish that devours the snails attached to the underside of a lily leaf. 'The bird stands on a leaf, commanding a small space of open water, and while the fish moves from one leaf to another it is snapped up' (36).

When at rest the Squacco appears to be uniformly-coloured brownish-buff, and is easily overlooked: but when flushed it seems to explode into conspicuousness, as it 'produces two white wings as if by magic' – only to vanish from view when, after a short flight, it settles again.

A small fish-eating species, with rather curious feeding behaviour, is the Black Heron (*Melanophoyx ardesiaca*) found most frequently in coastal regions of East Africa. When fishing this bird adopts an unusual posture, standing motionless in the shallows with the bill pointing downwards and the wings spread in a wide circle, so as to form a canopy, or tent, over the head. This habit may assist fishing in two ways: the bird's vision will be shielded from sun glare and reflected light, thus giving a clear view through the water's surface; and secondly, it has been suggested that the patch of shade thus produced may itself attract fish seeking shelter, and so act as a lure to prey.

Black Heron hooded while fishing (based on a photograph by Milon)

Another small species is the Green-backed Heron (*Butorides striatus*) found in dense, coastal mangrove swamps, and inland along tree-gir rivers and creeks. Being partly nocturnal, it hides up by day near the water; and has been observed 'sitting for hours immobile on some mangrove root, as if it realized how well its plumage harmonized with its surroundings'. In its mangrove habitat it feeds on crabs and on the fin-walking gobies or 'mud-skippers' (*Periophthalmus*).

The Black-headed Heron (*Ardea melanocephala*) is a terrestrial bird, much less dependent upon water for its food than other herons. When searching for its prey, which includes rodents, snakes, lizards, frogs and insects, it adopts the walk-stalk technique. Sir Frederick Jackson thus describes the movements of the bird when feeding in a sweet-potato plot: 'On arrival it always remained quite stationary for some minutes before commencing to move very quietly and cautiously, drawing up each foot slowly and extending it at each step with the utmost delibera-tion, with neck stiff and held well forward. On detecting its prey, rat, lizard or insect, it occasionally snapped it up at once; but more often it began to sway its neck sideways, slowly at first, but getting faster and faster until the body also began to sway. While this movement was going on, its head and neck were being gently lowered until within exact striking distance, when both were suddenly shot forward. In this manner it caught several small rats, striped mice, and lizards.' I have seen Black-winged Stilts make similar lateral movements of the head when about to strike at a fly – using parallax as a device for accurate range-finding. Nine Black-headed Herons which I examined from Jinja and the Lower Semliki Valley all contained the remains of insects: four had eaten rodents; and other prey recovered from their stomachs included Ranid frogs, a skink (*Mabuia*), a snake, a crab, and water bugs, dragonflies, moths, termites, grasshoppers and a spider. One of these birds had made a good meal of fourteen frogs!

Like the herons, the egret species show ecological differentiation in their feeding behaviour. The Great White Egret (*Casmerodius albus*) is a fisher, using the stand-wait-and-strike tactics. A solitary bird, conspicuous in its snow-white plumage, it adopts a familiar stance with the body inclined and the neck craned forward. Fish prey recovered from this species in Uganda included lungfish (*Protopterus aethiopicus*), catfish (*Clarias lazera*), various Cichlids (*Tilapia* and *Haplochromis*) and the air-breathing Anabantid *Ctenopoma muriei* – a fish tolerant of stagnant water and related to the Climbing Perch.

Unlike the last, the Little Egret (*Egretta garzetta*) hunts its prey actively, capturing crustacea, insects and small fry near the water's edge. When feeding it is constantly on the move, lunging and darting about, or wading along in the shallows to stab at prey in its path.

The Cattle Egret, or Buff-backed Heron (*Bulbulcus ibis*) is much the commonest member of the group. The birds are a conspicuous and familiar sight in game or cattle country where they accompany elephant, buffalo, eland, zebra, or domestic stock, feeding on cryptic prey that are forced into showing themselves when flushed by the grazing mammals. Grasshoppers are an important item in their food, which also includes toads (*Bufo*), tree-frogs, skinks (*Mabuia*), dragonflies, beetles, crickets, moths, termites and spiders.

Ecological divergence of species is again apparent when we study the feeding habits of storks. The Yellow-billed Stork, or Wood Ibis (*Ibis ibis*), is a handsome black and white bird, standing three feet in height – readily distinguished from the White Stork by its long, slightly decurved orange bill and bare bright-red skin of the face. This species frequents streams and water-holes as well as lake shores and sand-banks of rivers, sometimes in quite large parties. When feeding it walks in shallow water, probing the bottom with its bill held slightly open; and often it can be seen standing with one wing extended horizontally to shade the reflection of sunlight as it peers through the surface.

The White Stork (*Ciconia ciconia*), seen in large numbers in East Africa when on its trans-equatorial migration, is a terrestrial feeder, often found far from water, searching open country for grasshoppers. The birds are much attracted by bush fires, where they congregate in numbers, keeping ahead of the flames to intercept lizards, snakes,

Buffalo with Cattle Egrets

ground-nesting birds, rodents, locusts and other animals as they attempt to escape the fire.

Another trans-equatorial migrant is the Open-bill Stork (*Anastomus lamelligerus*). A specialist in its diet of molluscs, this bird searches shallows for slugs and water snails. Its bill is so shaped that when the mandibles are closed a gap remains between them, in which slippery gastropods are held and their shells broken, as with a nutcracker. Stomachs of two specimens I examined from Lake Victoria contained thirty-three and thirty molluscs respectively – mostly the water snails *Bellamya* and *Pila*. The open-bill's migration from the southern savanna, where it nests at

Marabou overlooking a crocodile's nest

121

the time of low water in July and August, to the northern savanna, enables the bird to profit by the reversal of seasons which thus provide optimum conditions for feeding throughout the year.

The Saddle-bill Stork (*Ephippiorhynchus senegalensis*) – the tallest member of the group, standing over four feet in height – is an elegant bird distinguished by its heavy, black-banded red bill which is surmounted with a yellow shield. This shy, solitary stork frequents the banks of

Marabou sailing

large rivers, swamps and lakes. Little has been recorded of its feeding habits. In the Victoria Nile at the time when crocodiles are hatching in late March and April, the Saddle-bill parades along the shore line near the rookeries, searching grasses, reeds and Nile cabbage in the shallows for hatchlings.

The Marabou (*Leptoptilos crumeniferus*) is a stork that has adopted a vulture's way of life. It is one of the first large birds the visitor to East Africa is likely to see, for each safari lodge in the game-viewing areas has its detachment of these birds which have become almost commensal with man – garbage-eating camp followers of the tourist. But this role is of recent origin. In a state of nature, Marabou consort with vultures at a kill. They also take live prey. Along the lower reaches of the Victoria Nile they are one of the most destructive enemies of the young crocodile. The drawing on page 121 illustrates a scene near one of my camping places where, at hatching time, Marabou would perch high in a dead tree overlooking a nesting ground. Sometimes, even before the eggs have hatched, the birds will probe deep into the sand with the long bill in search of embryos. In the absence of the female crocodile, half a dozen marabou will quickly account for a whole brood.

When not satisfying his appetite, the Marabou just stands about, giving one the ridiculous impression of his being deep in thought. He walks delicately, like Agag, but in a purposeful way, with the dignified, unhurried manner of a shopwalker. When seen at close quarters the bald pink and grey scabby head and sunken eye gives the bird a rather repulsive look, as of extreme old age in man. But on the wing this huge bird has grace and poise, as he sails in wide circles to gain height astride a thermal, his neck retracted, legs trailing, and his gular pouch flapping in the breeze.

The specialists: flamingo and skimmer

The flamingos are unique for their specialized mode of life and feeding habits. They inhabit the Rift Valley lakes, where they occur in gregarious multitudes – Lakes Hannington, Nakuru, Elmenteita and Natron being the most favoured. These are shallow, alkaline lakes with a high content of sodium carbonate and a rich growth of phytoplankton. Although found together in the same waters, the two African species have different feeding techniques and different diets, and so can co-exist without

The Lesser Flamingo (*Phoeniconaias minor*) seldom immerses the head when feeding, but skims the water surface of in-shore shallows, to collect the microscopic blue-green algae and diatoms which are its food. The Greater Flamingo (*Phoenicopterus ruber*) feeds farther from the edge, in one or two feet of water. Unlike its congener, it is a bottom feeder, immersing the head and moving forward with the bill pointing backwards and inverted in the mud. Its food consists mainly of midge larvae (Chironomidae), waterbugs (Corixidae) and copepods.

A highly specialized structure is the flamingo's bill, 'with the lower jaw large and trough-like and the upper small and lid-like'. With its associated parts it becomes a highly efficient filter – especially in *P. minor* whose habits require a straining mechanism of extremely fine mesh. The thick and fleshy tongue fits into a groove in the lower mandible and works to and fro like a piston, pumping water in and out three or four times a second. Hair-like lamellae which cover the inside of the mandibles strain out micro-organisms contained in the passing water. It has been calculated that when present in their maximum numbers Lesser Flamingo remove from Lake Nakuru every day about one hundred and fifty tons of algae.

Another unique, but utterly different, method of feeding from surface water has been perfected by the skimmers, or 'scissor-bills' – long winged, short-legged birds resembling terns. The family Rynchopidae contains but three species, found on coasts, lakes and larger rivers of the

African Skimmer (*Rhyncops flavirostris*): the bill from the side and from below, and the method of feeding in flight

warmer parts of America, India and Africa. *Rynchops flavirostris* is widespread in the Ethiopian region, but local in its East African distribution. Like the terns, to which they are related, skimmers are sociable and gregarious, sometimes occurring in dense flocks of hundreds, that rest by day on sandbanks. One such resting place in Uganda is the estuary of the Namsika sand-river a few miles above Paraa, where the black and white plumage and red bills of the massed birds provide a striking spectacle for visitors on their way to Kabalega Falls.

The curious shape of the bill (which cannot be used to pick up food) sets these birds apart. Both mandibles are laterally compressed – the lower one to a knife-like thinness. This lower blade juts out far beyond the tip of the upper mandible. When feeding the birds fly low over the water, with the bill held open at a downward angle and the lower mandible cutting the surface. In this way they scoop up food – possibly tiny fish or algae. They feed in flocks, mostly at dusk or by moonlight; and for this they require open stretches of calm water.

Many years ago I saw a company of these remarkable birds feeding at twilight on the Shire River in Mozambique. They approached fast, going in an extended line-abreast formation that nearly spanned the river from bank to bank. The dawn was dead calm, and as the flock swept by, thirty bills ploughed the water into parallel furrows. The skimmers had come in an instant, and were as soon gone from sight; but such a spectacle is not easily forgotten.

Diurnal birds of prey

The birds of prey are an extraordinarily successful group. Every habitat has its representatives; each species has its particular hunting technique; and every suitable prey-type has its falconiform enemy.

East Africa supports a rich and splendid raptor fauna, ranging in size from the Lappet-faced Vulture – a huge bird measuring forty-two inches from head to tail and with a wing spread of over nine feet, to the little Pygmy Falcon – about seven inches in length and weighing a mere two ounces. Over sixty species have been recorded from Kenya's Tsavo National Park alone, which is half as many again as the number recorded from Europe.

Between them the Falconiformes subject a wide range of possible prey to hunting pressure. Indeed, as Leslie Brown has remarked: 'Almost all animals, from elephants to small insects, are either eaten as carrion or preyed upon as live prey by some species of raptor'. The wide adaptive radiation achieved by the group is illustrated by reference to the following table which shows the main or favourite food items of various species. I have included in the list one – the Everglade Kite – not found in Africa, because of its interest as a specialist. A dedicated snail-eater, it is perhaps the only monophagous member of the order. Murphy tells us how, having taken its prey to a perch, 'It sits on one foot and holds the snail gently in the other, doing nothing that would inhibit the mollusk from emerging from the whorl of the shell. The bird makes no effort to obtain its food by force; it waits for the voluntary extension of the animal beyond the aperture. When that happens, the bird quickly

Species	Staple or preferred food
Crowned Hawk Eagle, *Stephanoaëtus coronatus*	Monkeys
Martial Eagle, *Polemaetus bellicosus*	Game birds, small antelopes
Saker, *Falco cherrug*	Duck
Lanner, *Falco biarmicus*	Medium-sized birds
Harrier Hawk, *Polyboroides typus*	Nestlings
Verreaux's Eagle, *Aquila verreauxi*	Rock Hyrax
Augur Buzzard, *Buteo rufofuscus*	Rodents
Bat Hawk, *Machaerhamphus alcinus*	Bats
Serpent Eagle, *Dryotriorchis spectabilis*	Snakes
Dark Chanting Goshawk, *Meliërax metabates*	Lizards
Osprey, *Pandion haliaetus*	Fish exclusively
Long-crested Hawk Eagle, *Lophaëtus occipitalis*	Frogs
Honey Buzzard, *Pernis apivorus*	Wasp and bee grubs
African Hobby, *Falco cuvieri*	Termites
Grasshopper Buzzard, *Butastur rufipennis*	Grasshoppers
Everglade Kite, *Rostrhamus sociabilis*	Snails exclusively (America)
Black Kite, *Milvus migrans*	Omnivorous
Tawny Eagle, *Aquila rapax*	Piratical on raptors
Egyptian Vulture, *Neophron percnopterus*	Offal, excrement
Lappet-faced Vulture, *Torgos tracheliotus*	Carrion
Lammergeyer, *Gypaëtus barbatus*	Klipspringer, bones
Palmnut Vulture, *Gypohierax angolensis*	Oil Palm kernel

pierces the snail behind the operculum, always in the same place which is evidently a nerve plexus. The kite sits and waits again, with the snail spiked on its maxilla... Gradually the muscles of the numbed snail relax. After two minutes, more or less, the kite vigorously shakes its head and swallows the mollusk whole, operculum and all, before the empty shell has reached the ground' (57).

The diurnal birds of prey locate their quarry almost entirely by sight, and capture it by surprise, agility and speed in the attack. Prowess in hunting and marvellously developed vision must have contributed much to the success of the group.

In basic structure, the vertebrate eye and a camera are similar. Both are optical instruments in which a lens produces an image which can be focused on a sensitive surface – the retina or film. If a photographer needs to record a wealth of critical detail, as for example in aerial photography, he uses (*a*) a lens of long focal length (which projects a large image on the film), (*b*) a camera having a large picture format (which receives a wide field of view), and (*c*) a fine-grain film (which will record critically defined detail).

Now the eye of a falcon admirably meets these requirements. The eyeball itself is as large as can be carried on the skull – in fact, the two orbits meet in the mid-line – and are together larger than the cranium. Thus the retina receives an image of large size and of wide angle. The retina itself and especially the sensitive fovea carries a dense concentration of visual cells, the 'cones', thus meeting the requirements of extreme visual acuity. In the Buzzard's fovea there are 1,000,000 cones to the square millimetre; in some hawks and eagles the eye's resolving power reaches a value of at least eight times that of the human eye. Another peculiarity of the falconiform eye is the possession of a second, or 'temporal' fovea which receives the image of objects in the forward line of flight. The temporal foveae enable the bird to see stereoscopically and thus to range-find accurately as it homes on to the target.

Hunting methods

Very marked divergence in hunting habits is found within the group. By exploiting available food animals by different methods and in diverse habitats, the pressure on prey species is spread, with a tendency to limit inter-specific competition between the predators.

Stooping. The method usually adopted by the Peregrine is to sight its quarry in the air, rise above it, and dive at terrific speed, the victim – often a pigeon – being struck with the hind claws. In the stoop the pointed wings are partly flexed, and the bird may reach over 200 m.p.h. in the full power dive. The Bat Hawk, a strictly crepuscular hunter, and a forest bird, adopts similar tactics to catch bats in the air. This bird has a very wide gape; and the bat is eaten from the claws during flight, and sometimes swallowed whole.

Still-hunting. The sudden pounce from a perch is practised by different species in both open country, and woodland. For example, the Pygmy Falcon watches from a bush – as do the shrikes – and drops on its prey in a gliding stoop, to return again to its perch. In wooded country the Little Sparrow Hawk and African Goshawk surprise their quarry with a swift sudden rush. Both have extreme manoeuvrability which makes them expert at negotiating cover when in pursuit at full speed. The Crowned Hawk Eagle is a forest species which takes its prey by the still-hunting method, dropping from above with a 'heavy swoop', after waiting perhaps for hours for its opportunity.

Aerial pursuit. The Hobby bears down on small birds in swift pursuit, and will engage in a follow-chase to overtake and kill such rapid fliers as martins, swallows and swifts.

Low level attack. Harriers quarter the ground, flying only a few feet up, and pounce on prey in the grass. Attack by low-level flight is of course another way to achieve surprise.

Hovering. The habit of the Kestrel, as the bird scans the ground for lizards and small mammals when poised in the air, is familiar. The Black-shouldered Kite also hunts by hover, dropping on its prey with the wings held over the back. This habit has also been perfected in another group of birds: the kingfishers typically dive from a perch, but the Pied Kingfisher hovers motionless far from land – a habit that enables the bird to fish open water inaccessible to other species.

Fishing. The Fish Eagle and Osprey are both fish eaters – the former mainly, the latter exclusively. The Osprey drops on its prey feet fore-most and fully extended, entering the water in a cloud of spray. The Fish Eagle descends in a stoop, but snatches the fish from the surface. In both species the scales beneath the powerful toes are modified into sharp spines, well adapted to grasp slippery prey.

Digging. The one bird of prey which can be said to dig for its food is the Honey Buzzard, which has the middle claw hollowed out as a scraper. Colonel R. Meinertzhagen observed a pair excavating a wasps' nest – tearing away grass and earth with one leg at a time; and catching the wasps as they came out. When the nest was reached, fifteen inches under the surface, the birds gorged themselves on pupae (53).

Foraging. The Harrier-Hawk forages in rock crevices, burrows, tree-trunks and birds' nests, for eggs and young birds, frogs, bats, mice or insects. Its long legs and small feet appear to be well suited for searching out hidden prey, and the bird can even bend its legs backwards as well as forwards at the intertarsal joint. In the branches of trees it can cling to bark in all positions as it searches for grubs. In West Africa it is described as an inveterate robber of nestling weavers – going from nest to nest and

24. Exerting full strength, this Lion dragged his buffalo-kill twenty yards into the shade of a bush

Young Lions feeding on a Buffalo on the shore of Lake Manyara

25. Sleeping Leopard rides a branch side-saddle

26. Two spotted cats with very different habits. The Serval is a nocturnal predator of bush country and forest

The Cheetah, fastest of land animals, frequents open steppe and semi-arid savaannas, and hunts in daylight

27. Hyaenas were feeding at this freshly-killed Zebra when a party of Wild Dogs ran up and drove them away

Spotted Hyaena; the heavily-spotted and well-groomed coat is unusual

28. The mother lies down to let the calf suckle

A long-horned Black Rhinoceros, Amboseli

29. White Rhinoceros at Umfolozi, Zululand

30. The newly-born calf is closely protected from crocodiles by its mother in the Victoria Nile

School of Hippopotamus in a wallow, Rwenzori

31. Frog-eating Hammerkops riding on a broad back

A battle-scarred Hippopotamus travelling at speed

32. Study of a bull Elephant browsing

33. Bull Elephant feeding on *Acacia tortilis* near Seronera

Cow Elephant with very young calf, Victoria Nile

34. The Skimmers rest, head to the wind, in a dense flock on a Nile sandbank

White-backed Vultures. Little but bone and hide is left of this carcass at Olduvai Gorge

35. Kori – the largest of the African bustards

Crowned Crane – Uganda's national emblem

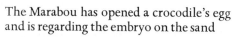

The Marabou has opened a crocodile's egg and is regarding the embryo on the sand

Marabou in the act of swallowing a baby crocodile

36. Black Terrapin (*Pelusios subniger*), Barotseland. This species is often eaten by crocodiles

Soft-shelled Turtle (*Trionyx triunguis*) laying eggs in a hole scooped out with the hind legs, Victoria Nile

37. Photograph taken in July 1952 of crocodiles on a former favourite basking ground at Magungu, lower Victoria Nile. By 1957 these animals had been exterminated by poachers and the rookery is now deserted

A huge crocodile rushes open-mouthed to attack a buffalo that attempted to swim across the Nile below Kabalega Falls

38. Waterbuck, when standing at the water's edge or swimming the river sometimes fall prey to crocodiles

39. A basking Crocodile looks across the river, while a Common Sandpiper takes an ectoparasite from the base of its tail

Nile Crocodile with jaws held widely agape. Gaping is a behavioural adaptation for the dissipation of heat

tearing them open with its bill; when so engaged it can hover like a humming bird without using its feet (2).

Walking. The Chanting Goshawk is a long-legged hawk that hunts for insects, on foot, in the grass. When hunting for frogs, the Long-crested Hawk Eagle walks slowly near water. But it is the Secretary Bird that has specialized as a walking hunter. Reptiles are an important prey, but the bird eats a variety of animals encountered in its path. A specimen shot near Cape Town had eaten no fewer than eight chameleons and twelve lizards belonging to two species, in addition to a tortoise, three frogs, an adder, two locusts and two quails. Bannerman records that on one occasion when a grass fire was raging, hundreds of snakes were destroyed, as they crossed the road, by Secretary Birds which had gathered for the occasion. In killing a snake the bird strikes with the pad of its foot, 'striking forwards and pounding its luckless victim to death. One realizes the force of the crushing blows delivered with the bird's foot once one has heard the noise of the stamping when a snake is being killed'.

Nut eating. The Palmnut Vulture has taken to a regular vegetable diet. Its favourite food is the kernel of the Oil Palm (*Elaeis guineensis*): the bird's distribution in East Africa nearly coincides with that of the palm tree.

Piracy. Several species depend partly for their food on piracy. The Bateleur, always on the wing and perhaps the most aerial of the Accipitridae, bullies and robs Ground Hornbills and Secretary Birds. The Fish Eagle is well-known to deprive Goliath Heron of their catch; it is also piratical against Osprey and pelicans and 'is always on the lookout to rob its neighbour'. The Tawny Eagle also robs other birds of prey and will sometimes drive vultures off carrion.

Dropping food. The Lammergeyer's habit of dropping food-objects from a height – to kill or to break them – has often been described. There are accounts of the bird knocking chamois, goats, and klipspringers off cliffs and then following to feed on the mutilated corpse. The Lammergeyer is also known to pick up tortoises, and drop them, to devour the contents of the cracked shell. Large bones, too, are dropped, and when broken the bird can lick out the marrow with its unusually long tongue.

Soaring and scavenging. Vultures watch the ground when soaring at great heights. It is said that they keep an eye on the jackal as a likely guide to carrion. They certainly watch and react to each others' movements.

Meinertzhagen has described how in Kenya he once watched from concealment beside a hartebeest he had killed. After seventeen minutes 'a single vulture came swishing down from the heavens . . . On looking up I could see birds planing down from all directions, some mere specks in the sky, others half-way down; after another ten minutes there were seventeen vultures on the ground . . . the latest arrivals must have come immense distances, drawn by seeing others descending'.

On the ground at a carcass the various species observe precedence, generally in order of size. The following sequence of events is typical. One morning when traversing Ngorongoro Crater, we came upon a congregation of vultures competing for food at a nearly-consumed Thomson's Gazelle. There were twelve White-backed and three Ruppell's Griffon Vultures gathered round, a Hooded Vulture in the background, and several Marabou standing by, waiting to snatch scraps torn off by the other birds. Though fewer in number, the Griffons dominated the scene. One or other would repeatedly threaten a White-backed Vulture, advancing with wings widespread and neck outstretched – like an attacking gander. The small Hooded Vulture is an opportunist, who gives way to others. As we watched, a huge Lappet-faced Vulture flew up and perched in a nearby acacia; whereupon all the vultures retired, even though the newcomer did not leave his perch.

Ruppell's Griffon Vultures

Scops Owl

Nocturnal birds of prey

Owls, like the diurnal raptors, show diversity in their feeding habits.
The beautiful, dark-eyed Verreaux's Eagle Owl – which can often be
seen hunting by floodlight at Kilaguni and Treetops – takes a variety of
prey including reptiles, game birds and poultry, mammals and insects.
But many other species have preferences for particular prey; and some,
like the fishing owls, are specialists. A few examples, listed on page 133,
indicate the extent of adaptive radiation found in the group.

Spotted Eagle Owl

132

Species	Preferred prey
Desert Eagle Owl, *Bubo ascalaphus*	Hares, jerboas
Cape Grass Owl, *Tyto capensis*	Rats, mice
Fraser's Eagle Owl, *B. poensis*	Bats
Spotted Eagle Owl, *B. africanus*	Small birds
Scops Owl, *Otus scops*	Insects
Wood Owl, *Strix woodfordii*	Beetles
Rufous Fishing Owl, *Scotopelia ussheri*	Fish

Remarkable adaptations in the owls are their powers of nocturnal vision, and the evolution of silent flight. What is it that makes the eye of an owl so extraordinarily efficient – enabling it 'to see in the dark'? On an earlier page we referred to the analogy between eye and camera. For photography in dim light, two essentials are (*a*) a lens which works at a wide aperture relative to its focal length, and (*b*) an ultra-fast film. Now, in terms of our analogy, the owl's eye meets both requirements. It has a lens and iris capable of working effectively at a very wide aperture – so admitting a maximum of light to reach the retina. And the light receptors of the retina, the 'rods', are highly sensitive to light of very low intensity. In addition, the forward-looking eyes of the owl give increased efficiency in dim light owing to the superimposition of two sets of images to form a single intensified conscious impression.

The astonishing powers of nocturnal vision have been studied in America by Lee R. Dice (20) whose experiments showed that the Barn Owl and other species can detect and pounce upon dead mice from at least six feet under an illumination of 0·000,000,73 foot candles: the intensity of starlight is about 0·000,08 foot candles – a brightness far above the owl's effective threshold.

Evolution of silent flight

A unique characteristic of the owls is their silence in flight. Flight of nearly all other birds is more or less noisy. Everyone is familiar with the startling clatter as a pigeon is flushed from cover. The larger forest hornbills fly with a loud swish of wings. A swooping falcon, heard at close quarters, makes the noise of a rocket. Even the smallest birds ride on whirring wings. In contrast, the phantom-like quality in the flight of owls is something to be wondered at.

The survival value for owls of silent flight is twofold. Not only does it enable a slow flier to surprise its prey almost until the fatal pounce, but also to locate prey by ear – without being distracted by the sound of its own wing-beats.

Silence is achieved by structural modifications of the feathers whose peculiarities have been investigated by Graham (31). Three silencing devices are involved. Firstly, the leading edge of the first primary feather is equipped with a fine comb-like fringe whose teeth jut forward and act as an air scoop, or series of slots, to smooth the flow of air over the upper wing surface. Secondly, the rear margins of the flight feathers have a fringe, like that of a Spanish shawl. These fringed trailing edges mute the mixing of upper and lower air streams so as to prevent noise-producing vortices behind the wing. Thirdly, all feathers have a soft, downy texture which serves to muffle sounds produced where adjacent feathers slide over one another during flight. Also, by comparison with most birds, the owl's weight is carried by a relatively large wing-surface; and this low wing-loading allows for buoyant, leisurely flight on muffled wings.

Very interesting is the fact that silencing adaptations are absent in the fishing owls – *Scotopelia* in Africa and *Ketupa* in Asia – whose underwater prey are unable to hear flight noise.

It is now known that the wing-beats of birds also produce ultra-sonic noise of frequencies that cannot be detected by the human ear. Since the majority of small mammals possess good hearing in the ultra-sonic range, this raises the question as to how far 'silent flight' of owls is also silent in the ultra-sonic range to which small mammals respond.

Tests carried out in Cambridge by Thorpe and Griffin (78), using sophisticated electronic equipment, showed that no ultra-sonic component was detectable in free-flying Scops, Little, Tawny, Barn and Long-eared Owls. On the other hand, further tests showed that fishing owls produced in flight ultra-sonic noise comparable to that in a flying Kestrel or Rock Dove. Thus in these aberrant owls, which have no need for the silent approach, wing silencers have either never evolved, or have been secondarily lost.

7
The Nile Crocodile

Hard by the lilied Nile I saw
A duskish river-dragon stretched along.
The brown habergeon of his limbs enamelled
With sanguine almandines and rainy pearl:
And on his back there lay a young one sleeping,
No bigger than a mouse; with eyes like beads,
And a small fragment of its speckled egg
Remaining on its harmless, pulpy snout;
A thing to laugh at, as it gaped to catch
The baulking, merry flies.

Thomas Lovell Beddoes

THE Nile Crocodile (*Crocodylus niloticus*) is one of three species found in Africa. The others, the Long-snouted (*C. cataphractus*) and the stumpy Broad-fronted or Dwarf Crocodile (*Osteolaemus tetraspis*) have a restricted range in West Africa. In historical times the Nile Crocodile – the largest

Crocodile sun-bathing

135

and best known – was abundant and widely distributed over the Continent, from the Nile delta to Cape Province and from the west coast to the Indian Ocean.

Exploitation and conservation

During the past twenty years its numbers have been drastically reduced in most parts of Africa where it was once plentiful. From important breeding places in East Africa such as Lake Victoria, Lake Kioga and Lake Mobutu the species has now almost disappeared. Its decline has two main causes. Firstly, the crocodile is affected by the general threat to all wild life which results from disturbance, and development of natural habitats. But today its survival is more directly threatened by uncontrolled commercial killing to meet insatiable demands of a lucrative leather trade.

This exploitation is to be deplored for several reasons. The animals represent a valuable asset in the countries where they are still found. Under rational management they could be utilized as a permanent source of high-grade leather. Moreover, the sight of crocodiles living in the wild state attracts visitors to their haunts like a magnet, and tourists bring currency to areas urgently needing trade.

On account of their role as master predator and scavenger, crocodiles play an important part in the ecology of inland waters. Furthermore the crocodilians are of immense scientific interest, being the only surviving members of the archosaurian group of reptiles – which included the dinosaurs and flying pterosaurs – that as Ruling Reptiles were the dominant terrestrial vertebrates throughout much of the Mesozoic. Crocodiles not unlike the modern forms were a flourishing group more than a hundred million years ago, and the study of the physiology, ecology and behaviour of species living today cannot fail to throw indirect light on the mode of life of their ancestors in past ages. Of particular interest is their reproductive behaviour. It is now known that the Nile Crocodile defends territory, that the pair have elaborate courtship displays, and that the female exhibits a degree of maternal care beyond anything found in other reptilian orders. Indeed, the highly developed behaviour patterns recall the breeding arrangements found among birds.

In addition to what has been said, these animals are surely worthy of conservation as creatures living their own lives, and in their own right 'as living monuments of almost unbelievable antiquity'. They chiefly deserve to survive – not because they happen to be useful or interesting to man – but for themselves, and for what they are.

The high walk

The crocodile's day

The diurnal movements of adult crocodiles follow a fairly constant pattern. The crocodile is nocturnally aquatic, and the haul-out to land begins in the hour before sunrise. Thereafter most of the day is spent basking ashore, though as the temperature rises there is considerable traffic into shade or back to the water. There are thus two main sun-basking periods, in the early morning and in late afternoon. Before sunset the return to the river has already begun and by dusk the grounds are once again deserted.

This diurnal cycle of activity is related to the reptile's thermal requirements. The term 'cold-blooded', or poikilothermic, is misleading

137

when applied to crocodiles in the tropics, for they have gone far in the direction towards the endothermal life of a bird or mammal. Cloacal temperatures recorded in East and Central Africa point to a surprising degree of thermal control – the mean temperature for specimens examined at different times of the day and night being 25·6°C (78°F), with observed fluctuations of only + 3·4°C and − 2·6°C. Thermoregulation is effected both by habitat selection and behavioural adaptations. Conspicuous among the latter, and most commonly seen during the hottest hours, is the habit of mouth gaping which promotes heat loss by evaporation from the mucosa.

When travelling on land the crocodile has three distinct gaits. In normal unhurried progression the legs are extended beneath the body,

The gallop

138

so carrying the belly high off the ground. This is the 'high walk', as distinct from the 'belly run' generally seen when crocodiles are escaping downhill, and at speed, to the water. At such times they toboggan on their smooth belly scales, using the legs as oars for propulsion. The third gait, very rarely seen and then only in young crocodiles, is the 'gallop', which resembles the bounding run of a squirrel. The crocodile swims with easy grace, moving like a huge fish, undulations of the powerful tail propelling the animal forward while the limbs are closely applied to the flanks.

Stomach stones

The stomach of an adult crocodile is always found to contain a quantity of stones. The occurrence of these stomach stones, or gastroliths, has long been known; and many writers have speculated on their function. Oldenburg, writing in 1665, says that the animal swallows stones 'for digestion'; and a century later Hasselquist compared the habit with that of seed-eating birds. The idea that the stones aid digestion by the trituration of food has since come to be widely accepted, though it is not supported by evidence.

The analogy with graminivorous birds is misleading because the crocodile is not a seed-eater and it has no gizzard. In early life – before stomach stones have been acquired – crocodiles feed upon prey with hard chitinous or calcareous exoskeletons, when trituration would appear to be most necessary; whereas the stone-carrying adults take fleshy foods that require no such treatment. Predatory fishes and fish-eating birds such as herons and pelicans have no gastroliths, though they sometimes feed on the same prey-species as the crocodile itself.

Examination of a large series of stomachs in East Africa and Zambia has shown that, while crocodiles in their first year never contain stones, adults almost invariably do so. Those living in a stony habitat acquire their quota earlier in life than inhabitants of meandering rivers with a sandy bottom, but even the crocodiles of stoneless swamps eventually manage to acquire their stony cargo. Such animals, living over a bottom of ooze and detritus, as in Bangweulu Swamp in Zambia, must of necessity make extensive journeys to collect their gastroliths. Crocodiles in St Lucia Bay, Zululand, have been found carrying water-worn pebbles that had been picked up from a beach miles away. Thus it is evident that stones are not ingested accidentally but are deliberately swallowed.

It is also found that while the absolute weight of the cargo increases progressively with age, adult crocodiles – whether from stony or stone-less terrain – tend to carry a standard load which amounts to about one per cent of the body weight. Maintenance of this weight-ratio suggests that gastroliths may subserve hydrostatic functions – a view that is supported by other considerations.

When a crocodile lies submerged, buoyancy reduces its weight by about 92·5 per cent. Thus a specimen weighing, say, 200 kg. will have a submerged weight of only 15 kg. Its 2 kg. of gastroliths will now account for about one seventh of its weight in water. Such a load will add significantly to a crocodile's effective weight when swimming. The load of stones carried by a large crocodile is considerable: a male from the Victoria Nile measuring 4·7 metres in length was found to have 4·77 kg. of cargo in his belly.

Such additional ballast is likely to be advantageous in two ways. Firstly, crocodiles have need of weight when struggling to hold large prey, such as a waterbuck or buffalo, under water until it is drowned. Additional ballast will also enable the animal to lie submerged on a river bottom – a habit often observed in rapids below Kabalega Falls – without being dragged by the current. A second and important function relates to the position of the stones in the body. Lying as they do against the ventral body-wall, they lower the animal's centre-of-gravity, and like the ballast in a ship's hold, act as a stabilizing force. The need for some such device is indicated by the instability of stoneless young crocodiles when placed in deep water, as compared with the easy poise of their stone-bearing elders.

Food, growth and sexual maturity

The crocodile leads a leisurely existence and, contrary to popular belief, takes relatively little food. In the choice of prey it is an opportunist, and its diet changes progressively with age – ranging from mosquitoes and ants to mammals as large as a full-grown buffalo. The young at first are wholly insectivorous, feeding on water-bugs, dragonfly nymphs, mole-crickets, beetles and spiders. At a later stage they take freshwater crabs, snails, toads and frogs, turtles, small birds and rodents. Fish become important prey in mid-life. Older crocodiles tend to feed increasingly upon reptiles and mammals.

At birth the hatchlings are about one foot in length. Growth is most rapid in early life – at a rate of about ten inches a year. But after the age of six the growth rate rapidly decreases, and older crocodiles yearly add only about an inch to their length.

The reptiles are surprisingly slow to attain sexual maturity. Males do not begin to breed until they are over nine feet in length, and females,

Bull crocodile on territory

about eight feet. Such animals are not less than seventeen years of age. In the pages which follow a brief account is given of successive phases of the reproductive cycle.

Territory

When working on Central Island in Lake Rudolf – one of the few remaining places in Africa where crocodiles can still be seen in the primeval

141

state, undisturbed by man – M. L. Modha (54) was able to record, for the first time, details of territorial behaviour. Observations at one of the island's crater lakes, about a quarter of a square-mile in area, showed that the entire shore-line was shared out by a dozen large male crocodiles. Each was in possession of a particular stretch of shore which it guarded by patrolling up and down, stopping and turning back at each end of its beat. Males intruding on a neighbour's territory were chased off by the rightful owner. Fights between dominant males were sometimes seen, but in general territorial rights were respected and actual combat avoided: 'Within his territory the male has absolute right of way and is hardly ever challenged.'

Modha found that individual territories ranged from 60 to 230 metres in length. On the Victoria Nile below Kabalega Falls, where recently I had an opportunity to study the pre-mating activity, dominant males were more widely spaced along the shore, though in certain secluded lagoons 'bulls' patrolled and defended shallows only some 90 by 50 metres in area. These display areas were already occupied in mid-November – more than six weeks before the egg-laying season.

During the mating time basking bulls often utter a bark or cough, the note being deep, loud, hollow and abrupt. But the most remarkable vocal demonstration is the full roar, which I have described elsewhere (14): 'The animal, which has been lying on the mud with jaws agape, first elevates its head and opens its jaws yet wider in what appears to be a prodigious yawn. The roar is a growling rumble, very deep in pitch, rattling, vibrant and sonorous, like distant thunder or the roll of a big drum, which is protracted and may persist for six or seven seconds. Jobson gives an apt comment when he speaks of the sound as if coming from the bottom of a well.' The biological function of the roar is not clearly understood: it appears to be a mating call directed towards members of the opposite sex, rather than a threat directed against rival males.

Courtship display

The courtship display has several phases. It is both dramatic and long-continued. Early in the season a male may be seen displaying, day after day, to one or more females that fail to respond. The male is poly-gamous; and the count of 58 dominant bulls, and subsequently of 152 nests in the same study area – between Paraa and Kabalega Falls – gives a ratio of 2·6 inseminated females to one male. It is interesting to note that at the Samut Prakan Crocodile Farm in Thailand, the sex-ratio maintained in the breeding ponds is one male to three females; but in the breeding season the male apparently forms a relationship with only two females.

When a female enters an occupied territory the bull may cruise up towards her. Then, sometimes after a long pause during which the pair lie almost submerged, he slowly lifts his head until the slightly parted jaws are clear of the water. At the same time he raises his tail to form a semi-circle, with its tip drooping behind into the water. This startling posture causes the whole body to submerge. His neck swells up. Streams of bubbles are now seen to emerge from the gullet. The body appears to tremble or vibrate violently, and bubbles and spray dance on the seething surface. This is the 'bubble display'. It is often followed by an even more spectacular 'splash display' – the jaws, lifted yet higher, are now brought down with a tremendous blow, or series of blows on to the surface. Often such a performance ends with a 'thrashing display': the tail is lashed from side to side, the jaws are champed with the sound of planks coming together, and the animal is wrapped in the turbulent commotion of water.

A variation of these extraordinary exhibitions is seen when the bull inhales to the full capacity of his lungs, and then takes up a position with his head inclined downwards and his snout submerged. He now forcibly expels air through his nostrils, and throws up a vertical jet or spout of water. The 'fountain display' may continue for five or six seconds. Perhaps the author of the Book of Job had himself seen this phase of courtship when he wrote: 'Out of his nostrils goeth smoke, as out of a seething pot or caldron.'

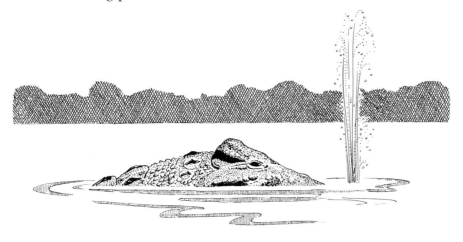

The fountain display

It will be noted that in all these various postures and actions, whose function is to excite the female, the water itself is made to contribute so largely both to the visible and audible components of the display. After displaying the male may approach the female; but frequently the female cruises away, and she is not pursued. Indeed, only occasionally does courtship result in copulation. During the demonstration the female's role is passive. But when ready to mate she enacts a prenuptial display: rearing her head and fore-quarters out of the water, she points her gaping jaws to the sky, and utters a creaking, guttural sound. Mating takes place in shallow water.

Prenuptial display of female

Hippopotamus shifting crocodiles

The nest

After insemination there begins for the female a prolonged period of maternal duties that will keep her occupied for six months. Choice of a nest-site is influenced by a number of requirements: a sufficient depth of soil to provide a pit where the eggs can be laid; nearby shade into which the guardian mother can retire during the heat of the day; and access to permanent water. Nests are usually found on river-side flats, low-lying islets, lake-side beaches, sand-spits or in the bed of a dried-up watercourse. Sometimes, where the animals have been much disturbed, nests are found on steep river banks or on top of inland cliffs twenty or even forty feet above the water. In rocky habitats where deep soil is not available, crocodiles will lay in shallow gravel beds and in such places the upper eggs of a clutch may even lie exposed to view. A crocodile will sometimes go four hundred yards from water to reach a suitable nesting place.

The female excavates an oblique burrow which leads to a spacious egg-chamber. Digging is done with the fore-limbs, the hind-limbs being

Females guarding nests at Namsika

used to shovel the spoil away from the burrow's mouth. Nests vary in vertical depth from about eight to eighteen inches. Between twenty-five to seventy-five or more eggs go to the clutch, the usual number in the Victoria Nile rookeries being about fifty-four. Larger clutches of more than a hundred eggs are sometimes found, but these are likely to be the product of two females. After laying, the crocodile fills in the cavity, rams the soil tight, and leaves the surface flush, so that the place appears never to have been disturbed.

In remote spots where the reptiles have formerly been free from human interference and allowed to breed as they have doubtless done from time immemorial, they nest in colonies, the nests lying so close together that, after hatching time, the rims of the craters are almost contiguous. Two such rookeries were known to me in 1952 on the southern shore of Lake Mobutu, where seventeen and twenty-four nests were crowded together in areas measuring 25 by 22 and 26 by 24 yards, respectively. In the thirties Capt. C. R. S. Pitman knew many such communal grounds on islands in north Lake Victoria. But colonial nesting is incompatible with present-day harrying and exploitation; and these ancestral rookeries have long since been deserted.

There is a fairly well-defined laying season which, however, differs according to locality. In Zambia and Barotseland new-laid eggs are found in late August and early September; on Lake Malawi about a month later; on Lake Mobutu and the Victoria Nile during late December and January; and on the White Nile in April and May. In each of these regions laying is, in fact, a dry-season phenomenon and occurs after the rains and when the waters are receding. The situation in northern Lake Victoria is anomalous, there being two laying periods; but these again correspond to the two annual periods of falling water level.

It is not yet understood how this synchronization is brought about. But its biological significance is evident: for the long incubation period of three months coincides with the phase of lowest water, when the nests are in least danger from flooding; while hatching takes place after commencement of the rains, when rising water facilitates dispersal of the hatchlings into flooded retreats at a time when a rich harvest of insect food will be available.

Defence of the eggs

The mother remains in attendance, at or near the nest, during the three-months incubation period – either lying over the nest with the throat directly above the eggs, or in nearby cover from which she has

Female crocodile in shade

the nest in view. Watch is kept by night as well as by day; and it appears, from stomach examinations, that brooding females fast. During this vigil – and in my experience especially when nesting at some distance from the water – the females become torpid and are most reluctant to leave the grounds even under the greatest provocation. Such a comatose female is shown above.

Egg-eating enemies take a heavy toll of unguarded clutches. The most persistent is the Nile Monitor. These large lizards are squatters on the rookeries, living in deep holes which they excavate nearby, and spending much of the day exploring and probing in search of eggs. Once a nest has been opened avian camp-followers, including Black Kite, Palmnut Vulture and Marabou, are attracted to the feast. Another important diurnal marauder is the Olive Baboon. Between them, monitors and baboons destroy many clutches; and at night the work is taken over by other raiders – Spotted Hyaena, Ratel and White-tailed Mongoose.

Defence of the eggs by the crocodile is mainly passive, each female lying over or near her clutch. But the reptiles have often been seen to make active sorties against predators, in a ponderous attempt to drive them from the rookery; and the presence of monitors in crocodiles' stomachs indicates that the egg thieves are sometimes caught unawares.

Maternal care at hatching time

At the depth at which the eggs are laid incubation proceeds steadily in a nearly constant temperature of about 31°C (88°F). When the young are ready to hatch they are sensitive to vibration and to air-borne sounds, and react by calling from the shell. If crocodile eggs are stored in a room, the tread of someone passing, the closing of a door, a cough or the human voice will elicit the yapping or croaking call. The sound is a short 'ao', not unlike the familiar 'Kyow' of the Jackdaw. In the field, the pat of a hand on the nesting place is sufficient to start a muffled subterranean chorus; and this response provides a ready means of ascertaining the time when hatching is due.

Under natural conditions the step of the mother or the sweep of her tail near the nest will have a similar effect. The sounds uttered by the unborn young stimulate the parent to open the nest, thus enabling the hatchlings to escape from the shell. The young are absolutely dependent upon maternal assistance at this time, because after three months of trampling the soil over the eggs becomes compacted. If the mother is not at hand, the young die in the brood chamber, being unable to force their way to the surface. At nests that have been deserted calling may continue for several days; and if then released the overdue hatchlings burst from the shell almost explosively as soon as the eggs are unearthed, though a moment before the shells were not even chipped.

Experiments carried out on different rookeries below Kabalega Falls in 1968 and 1969 have shown that crocodiles can be summoned from a distance by a tape-recording of the cries of unhatched babies. Remarkable determination of the female to reach and free her offspring has been observed on a number of occasions. This was first demonstrated by Voeltskow (80), who had a nest surrounded with a fence. After the female had several times broken through, the fence was replaced by a stronger one. The female then dug a deep ditch in her efforts to reach the nest. In Zululand A. C. Pooley (62) recently isolated a nest with a structure of stout poles and 8-gauge wire: at hatching time the mother smashed her way through the barrier. Screens of heavy wire mesh, laid over nests to protect clutches from predators, were tossed aside and left in a twisted heap of wire and pegs.

Brood care

One might expect that, on hatching, the young would make for the nearest water, as turtles do, running the gauntlet of predators on the way. But this is not their habit. Instead, as soon as they have struggled clear of the shell they crawl towards a nearby patch of shade and creep into any crevices which afford some concealment. And in their search for cover they often head inland and away from the water which will later be their home. Hatchlings will even run to a man who happens to be standing by the nest, and climb on to his shoes. During the first hours of life members of the brood are in fact reluctant to enter water; if placed in deep water they hasten ashore, to hide in reeds, tussocks or any convenient cover. How are we to explain this behaviour?

During surveys of nesting crocodiles carried out on the Victoria Nile in 1969 and 1972, I placed a dummy of an adult crocodile near nests at the time of hatching, so as to simulate conditions at an undisturbed site. Under such circumstances most of the young – sometimes within seconds of emergence and while still attached by membranes to the shell – would hurry to their 'mother' and cluster together under cover of her throat and flanks.

It is known that the female conveys her brood from the nest to a secluded nursery site. This may involve a journey of several hundred yards when the nest is in a dry sand-river or, as sometimes, when the selected site for the nursery is an isolated water hole. The transit usually takes place in the early hours before dawn; and details of this critical phase of the breeding cycle have never been observed. How is this journey accomplished? Are the young conducted, or are they carried to their destination? One of the Paraa boatmen with long river experience assured me that he had himself seen a crocodile on land carrying a brood of newly-hatched young. In Nigeria Lamborn (42) was told by natives that the hatchlings attach themselves to the dorsal fringe of the

Crocodile hatchlings in nursery

mother's tail. Natives of Lake Rudolf told Modha that the female carried her young on her head, neck and back.

It would be too easy to dismiss these tales as myths. However, I tried the experiment of slowly dragging one of the above-mentioned dummies away from a nest where the young had just emerged. As soon as their 'mother' began to move, hatchlings that had been sheltering beside her followed in haste, mounted her tail-region and clung to it.

Life in the nursery

The nurseries are established in places where there is slack water and weed cover, for example, in the lee of a fallen tree and floating vegetation, in small weed-choked riverside ponds, or in water-filled creeks. Observation at many nurseries has shown that they are occupied continuously for several weeks and in some cases up to at least three months before the young disperse. During this time they remain tightly gregarious and are closely guarded by the female both by day and night. If, at a sudden fright, the pack becomes scattered, the hatchlings communicate by croaking calls and soon reassemble close to the parent. The young often use the partly submerged mother's body as a basking place and may be seen crowding on to her head and shoulders (Plate 48).

Unguarded hatchlings are very vulnerable and are preyed upon by many enemies, including Saddle-bill Stork, Great White Egret, Marabou, Goliath Heron, Fish Eagle, Black Kite, Palmnut Vulture and Ground Hornbill. These, together with monitor lizards, search for them along the waterfront. When in charge of her brood, the female becomes an aggressive and dangerous animal and will even attack a patrol boat, or a man on land. This is a striking reversal of normal behaviour, for at other times a crocodile's almost invariable reaction to human beings – when surprised ashore – is to escape to the water. One female, whose brood was in a narrow ditch fifty yards from the river, made repeated attacks when I approached closely from inland to observe and photograph – hissing, growling, making lunges, jaws agape, with incredible agility and sending spray flying with tail strokes. Seen at point-blank range such demonstrations are most impressive.

8
Concealing Coloration
and Disguise

Themselves but shadows of a shadow world.
Tennyson

ONE of the fundamental facts affecting the lives of wild animals is the interspecific struggle for existence. The problem of self-preservation is very real, very urgent, and often difficult enough to solve. But it is one with which all forms of animal life are faced. Individual survival depends upon the satisfaction of two primary needs – food and safety. In a world peopled with potential enemies, and pregnant with hunger and the possibility of starvation, if an animal is to survive it must eat and avoid being eaten.

The urgent nature of this central problem of self-preservation is reflected in the diversity and specialization of evolutionary experiments in adaptation. We see evidence of this in the development of speed, on land, in the air, and under water, by pursuer and pursued; in the use of stealth and surprise, of deception and ambush; in the display of warning signals, or of alluring baits; in the elaboration of smoke screens, traps, nets and parachutes; in retreat, obtained by burrowing underground, by taking to the branches, or by the adoption of nocturnal habits; in protection afforded by plated or spiny armour; in the development of poison, and of deadly apparatus in the form of fangs or stings for its injection into the bodies of enemies or prey; in the principles underlying radar, put to practical use by the Oil Bird, and by bats and whales; and in chemical warfare, notably practised by certain insects.

Of all these various adaptations – which it will be noted each has its parallel in the paraphernalia of modern war – perhaps none is so important, so widely distributed, or so perfect as that which renders animals inconspicuous, and often well-nigh invisible in their natural surroundings. It is not too much to say that concealment appears to have been one of the main ends attained in the evolution of animals.

Recognition and concealment

Before considering the methods by which different animals are rendered inconspicuous and often extraordinarily difficult to recognize in the field, we must bear in mind the optical principles upon which recognition depends. When we look at any object, it appears in the field of vision as a patch of colour, more or less differentiated from surrounding objects which form its background – in colour and tone, in light and shade, in surface and contour, and in the shape of its shadow. Visible form can only be distinguished when it exhibits one or more of these clues to recognition. With the reduction of such clues, an animal or any other object becomes more and more difficult to detect; in their absence it becomes unrecognizable.

It follows that four fundamental steps towards effective camouflage must lie in the direction: of *colour resemblance* – the agreement in colour between an object and the background against which it is seen; of *obliterative shading* – counter lightening and darkening which abolishes the appearance of roundness, or relief, due to the effects of light and shade; of *disruptive coloration* – the use of a superimposed pattern serving to break up surface continuity and to blur the outline: and of *shadow elimination* – the screening or effacement of cast shadows by orientation, or by structural adaptations. It is very interesting to note that these theoretical principles of colour resemblance, countershading, disruption and shadow concealment – together with appropriate patterns of behaviour – are those found to operate in nature, whereby different animals, belonging to the most diverse groups and living in the most dissimilar surroundings, are often rendered so extraordinarily difficult to detect when encountered in the field.

General colour resemblance

The general colour resemblance borne by many animals to the surroundings in which they live is a theme familiar to everyone. Every major environment with a dominant type of coloration furnishes innumerable examples of the principle – different members of the fauna wearing a cryptic uniform: white in the snowlands; ochre or buff in the desert; green among the evergreen foliage of tropical forests; dull and dappled beneath the trees; striped among grass and reeds; bluish, or transparent, in the surface waters of the sea.

It is interesting to notice that in a given environment similar cryptic colours have been developed independently in the most distantly related groups of animals, while the coloration is itself due to a variety of causes – chemical, physical and physiological. For instance, among foliage dwellers, green coloration has been adopted as a uniform by many families of snakes and lizards, tree-frogs and birds, as well as by innumerable species of caterpillers, grasshoppers, cockroaches, moths, mantids, beetles, bugs and other insects; and it is found in similar environments in South America and Australia, in tropical Asia and tropical Africa.

In most birds green is a structural colour, being caused by the scattering of light in minute air-filled cavities of the feather. When soaked in a downpour the parrot's fine dress turns to subfusc, until such time as the feathers have dried again. Unique among birds is the green coloration found in members of the Musophagidae, such as Fischer's Turaco and the White-crested Turaco; this is due to a green pigment, appropriately named *turacoverdin*. Improbable as it may seem, the bright leaf-green tints of tree-frogs like *Leptopelis* is brought about by the arrangement of three kinds of pigment cells in the skin – melanophores, lipophores, and guanophores, whose pigments are black, orange and white, respectively. Again, insects have discovered many different ways of producing green coloration. In some caterpillars the chewed fragments of leaves in the gut show through the transparent body wall. Other insects synthesize blue-green and yellow pigments, which when present in the blood are mixed like water-colour paint, but when deposited in the epidermis, as in stick-insects and grasshoppers, are scattered in separate granules to produce green after the manner of an impressionist painter.

Adaptive behaviour

Coloration alone cannot conceal an animal. Adaptive behaviour is of vital importance in rendering the colour-scheme effective, in other words, for survival. Movement, in particular, is at once detected: hence the staff sergeant's warning to cadets on the parade ground: 'Right or wrong, stand still!'

When danger threatens the Bittern assumes a characteristic and grotesque pose, becoming lost to view against its background of reeds, and remains inanimate as in a trance. Young of many plovers and stone curlews lie flat and motionless while the intruder passes the nesting place. Crouching in the short grass of the Serengeti plain the African Hare will instinctively freeze, and allow the closest approach of a vehicle without stirring.

This self-preserving instinct is highly developed in the fawns of Thomson's and Grant's Gazelle. Unlike the young of hartebeest and wildebeest which follow the mother from birth, gazelle fawns lie alone, often for several hours, while the mothers go away to graze. Walther (81) found that this behaviour dominates their life for the first two weeks: 'While staying put in this manner the fawns almost disappear from the face of the earth.' When the mother returns to the fawn she does not go to the place where it is lying up, but calls it to her; and ideally, the fawn does not move until the mother returns. She nurses and licks it. The fawn then goes off to lie concealed, often at a new site, while the mother watches from a safe distance.

For the first few days a fawn will lie still after discovery, even to the point of allowing itself to be handled. So strong is this instinct that during a grass fire in Ngorongoro Crater in September 1963 several hidden fawns of Grant's Gazelle 'were literally plucked from the flames' (23). When trying to evade predators, even adult gazelles will occasionally resort to lying motionless, with head and neck stretched out along the ground. Dr Kruuk once watched a female Thomson's Gazelle being hard pressed by wild dogs in open country. As she passed a patch of high grass, she dropped and lay still: several of the dogs stopped short, while others passed on without finding their quarry. On two occasions Walther saw male 'tommies' escape in this way from wild dogs that had been in full chase.

Clearly defined habitat preferences are shown by various African larks for soils of different colours. Not only do various species and races of *Ammomanes* and *Mirafra* resemble the particular ground on which they live – whether black lava, reddish earth, or white gypsum sand; but also, the birds are extremely reluctant to leave their own terrain and cannot be driven onto adjoining ground of a different colour.

Colour change

Many animals possess the power of colour change, being able to assimilate their colour to that of their background. The nature of such changes differs according to circumstances. Some are permanent, occurring once in the life history. Others are temporary and capable of repetition. Some changes are built up gradually by the increased production of some particular pigment; others take place quickly, in minutes, in seconds or in an instant, by the movement of pigment in epidermal cells – the chromatophores.

Locusts and other grasshoppers possess powers of colour adjustment within wide limits, from whitish to black according to the colour of their environment. On green vegetation, as in the Serengeti short-grass plains, grasshoppers are bright green. In March and April these insects form the main food of migrating Abdim's and White Storks. But on ground that has been blackened by a grass fire, grasshoppers and other insects become darkened with black melanin pigment deposited in the cuticle, and look like burnt straw or charred chips of wood. Caterpillars taken on an island in Lake Victoria by Hale Carpenter were coloured in longitudinal stripes of coal black and bright green, and so harmonized perfectly with fire-blackened grass stems and new green shoots where they were feeding.

Obliterative shading

The second optical principle upon which concealing coloration depends is that of countershading. Owing to the effect of unequal illumination falling upon its different surfaces, a solid object of uniform colour presents to the eye the well-known appearance of light and shade, or 'relief', to which is due its appearance of solidity. By this means alone an object can be distinguished as a solid form – even when it is placed before a background whose colour and texture exactly matches its own.

When an animal or any other solid body is observed out of doors in the open, it will be seen that its upper surface is more brightly illuminated than its under parts, owing to the direction of incident light from the sky. The effect of this top lighting is to lighten the tone of the upper parts while the lower surfaces which are in shade appear to be darkened.

By countershading the upper surfaces, and counterlightening those beneath, using properly graded tones, it is possible to nullify the effects of light and shade, and thus to render a rounded body apparently flat. It will be noted that such a result is brought about by the use of tones that are darkest above, becoming gradually lighter on the sides, and lightest beneath.

Now, as was pointed out many years ago by Abbott H. Thayer, the American artist with whose name the principle of countershading will always be associated, this arrangement of graded tones forms the basic colour scheme of innumerable wild animals. Once understood, we see examples everywhere. The brush of nature has laid down in skin and scale, in fur and feather, darker pigments on the back, graded into paler pigments on the belly.

In countershading we have a system of coloration the very opposite of that upon which an artist depends. The artist, by the skilful use of light and shade, creates upon a flat surface the illusion of solidity; nature, on the other hand, by means of countershading, creates upon a rounded surface the illusion of flatness. The one makes something unreal recognizable; the other makes something real unrecognizable.

Many East African mammals – such as Impala, Reedbuck, Oribi, Lesser Kudu, dikdiks, duikers, Klipspringer, hyraxes and the African Hare – owe their inconspicuousness in the field to this system of coloration. It is particularly effective under diffused lighting, when the sky is overcast; and on a moonless night such animals appear ghost-like, as insubstantial objects. The colour photograph of a Kirk's Dik-dik provides a very striking example of illusory flattening of the body that results from obliterative shading (Plate 40).

The occurrence of obliterative shading is closely correlated with the habits of the animals concerned. Thus it is of particular interest to notice that cryptic insects which habitually rest in a back-down position, have the normal colour scheme also turned upside-down: that is to say – they are coloured darkest on the belly and lightest on the back. A beautiful example is furnished by the larva of the Eyed Hawk Moth: here both posture and coloration being inverted, the stout green caterpillar becomes optically flat, like the surface of a leaf.

Larva of Eyed Hawk-moth. *Left*, natural (inverted) resting attitude, showing obliterative effect of countershading. *Right*, unnatural attitude, showing strong relief and solid appearance

The nest

Disruptive coloration

It will be clear from what has already been said that, under ideal conditions, colour harmony combined with countershading are alone sufficient to render an object virtually invisible. But conditions change, and animals are seen against a constantly varying background, and light which changes in intensity and direction. Objects therefore tend to stand out as continuous surfaces of colour, differing more or less from their background, and outlined by characteristic shapes. It is this con-

159

tinuity of surface, bounded by a specific contour, which provides the chief clue to recognition.

The difficulty is met by the application of patterns, such as those exhibited by African Rock Python, Gaboon Viper, Puff Adder; by chameleons, geckoes and frogs; by larks, francolins, nightjars and the eggs and nidifugous young of plovers, stone-curlews, jacanas, waders and other birds; by giraffes and spotted cats; and by innumerable species of grasshoppers, mantids, moths and other insects. The function of such patterns is to prevent, or to delay, the first recognition by sight. The patches of contrasted colours and tones with which such animals are marked tend to catch the observer's eye, to distract his attention from the underlying form upon which they are displayed, and to be accepted as incidents of the background. The extraordinary effectiveness of even simple patterns, *when combined with countershading*, as a means of delaying or baffling observation, is well shown in the photograph of a male and

The Horned Escuerzo, Amazon

Masai Giraffe and fever trees

female Lesser Kudu (Plate 49). Here vertical white stripes are super-imposed on the countershaded body, and the animals seem to become part of their background.

The effect of disruption is thus to break up what is really a continuous surface into what appear to be a number of discontinuous surfaces, which tend to be interpreted by an observer as separate objects. This visual device is carried a stage further in the case of many animals in which elements of the pattern act in an opposite way, serving to join together surfaces which are in fact discontinuous. Such 'coincident' patterns which sweep without interruption across various parts of the body – legs, fins, wings or eyes – act like masks.

The principle is well illustrated in the colour scheme of a small East African tree-frog *Megalixalus fornasinii*. This animal bears on its back a pair of broad, conspicuous, silver-white stripes. Similar stripes occur on the hind limbs. These markings are so disposed, that in the normal

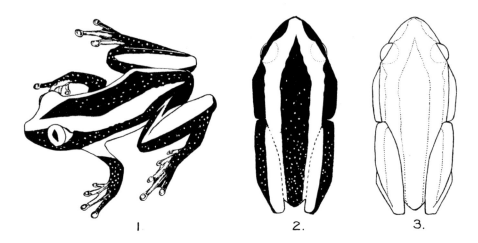

The tree-frog *Megalixalus fornasinii*

resting attitude (when the limbs are closely applied to the sides of the body) the stripes on the back coincide with, and become a continuation of, those on the legs. The coloration and attitude combine to produce an effect whose deceptive appearance depends upon the breaking up of the entire form into two strongly contrasted areas of colour, neither area resembling part of a frog.

Similar coincident patterns occur widely in nature, and are used to span the spaces between the folded segments of the leg in many frogs;

between the wings and body in grasshoppers; between the four out-spread wings in many cryptic moths; or between upper and lower jaw in various lizards and snakes. Such patterns are very widely used in the camouflage of the eye itself. Here the method – employed by many fishes, frogs, snakes, birds and other animals – is to include the black pupil within a stripe or band of black pigment, which sweeps across the iris so that the eye becomes incorporated and lost within the general pattern of the body. An example is the vine-like tree-snake *Oxybelis acuminatus* from tropical America. The most perfectly developed ocular masks are found in various reef fishes of the Indian Ocean. Some species, having effectively camouflaged the real eye, have taken deception a stage further by displaying a very large and intensely conspicuous dummy eye, or 'ocellus' near the tail.

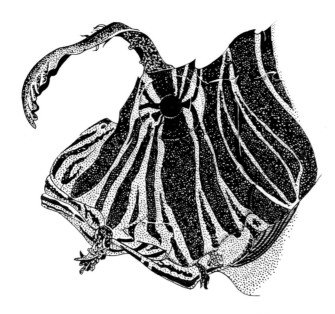

Disruptive eye-masks: *above, Oxybelis acuminatus* and *below, Pterois volitans*

Concealment of the contour and shadow

Pattern can be used to break up a continuous contour just as it can to break up a continuous surface. Contour obliteration is induced when contrasted elements of the pattern cut across, and are interrupted at the margin. Conversely, the contour is emphasized when the pattern conforms to it. The principle can at once be appreciated by comparing similar objects bearing the two types of pattern. If the rectangles in the

Diagram illustrating the use of a disruptive pattern in the obliteration of contour

figure are viewed from successively increasing distances, it will be found that the border of the right-hand figure will blend with the background at a point from which the other figure is still plainly visible.

Interrupted patterns that break against the contour are found in many classes of animals. Familiar examples are giraffe and zebra, whose patterns of contrasted dark and light tones tend to mask the animal's outline. In full sunlight or in open country a zebra may appear conspicuous; but in the dusk or under a night sky the combination of countershading and disruption is wonderfully self-effacing.

Tell-tale shadows cast by an animal on its surroundings are obliterated by various devices – behavioural and structural. For example, various Satyrid butterflies which habitually rest on the ground, with the wings closed over the back, orientate the body in relation to the sun on alighting, so that the shadow cast by the wings is reduced to a mere inconspicuous line. In this position the insects are extremely difficult to detect, for the exposed under-surface of the wings bears a cryptic design which closely harmonizes with the surroundings.

Other animals meet the difficulty by crouching flat when exposed

40. Gerenuk, or Waller's Gazelle, Samburu

Kirk's Dik-dik, Lake Manyara National Park

41. Red Elephant, Tsavo National Park, Kenya

Black-maned Lion, near Keekerok, Kenya

to danger, with the body, wings or other parts closely applied to the surface on which they are resting. This arrangement is well illustrated by the young Stone Curlew, and by many cryptic ground-nesting birds when incubating. The female Ostrich, when attending the nest by day, flattens her head and neck on the ground.

Even among mammals, similar tactics are sometimes adopted. F. C. Selous, the famous hunter and explorer, observed that the smaller African antelopes – Steinbuck, duikers, Oribi and Reedbuck – will, while keeping their eyes fixed on the unfamiliar object, crouch slowly down, and then, with their necks stretched along the ground, lie waiting. Vaughan Kirby noted similar behaviour in the Bushbuck. His book *In Haunts of Wild Game*, written in 1896, has the following account: 'A particularly fine bushbuck had evidently been walking along the bank, or perhaps drinking at the river, as I came along, and had seen me, and at once lain down close to the bank, amongst the stones, with his head stretched out along the ground, and his horns consequently pressed flat back upon his neck. Crouching thus amongst the dark stones, the tops of which were rounded, and just about the height of his back as he lay down, his colour – darkened by the rain which was falling heavily – so exactly assimilated with that of the grey rocks around that he was practically secure from detection.'

The same principle is seen in certain tropical geckoes, such as the Malagasy bark gecko *Uroplatus fimbriatus*, where the squatting habit – in this case on the bark of a tree – is greatly enhanced by flap-like out-growths which extend from the sides of the body and tail, and serve both to screen the shadow and to join the body to the tree trunk which it closely resembles, so that the animal appears to form part of its natural resting place.

Background picturing

The last example leads us to another aspect of concealment, which may be considered under the heading 'background picturing'. A further step towards effective camouflage is taken when an animal wears a cryptic pattern which more or less closely reproduces the background-pattern against which it is normally seen – as happens, for instance, in the case of many cryptic Sphingid and Geometrid moths, and bottom-dwelling fishes. Such animals reproduce on their bodies a replica of their normal habitat – of bark or lichen or gravel. These colour schemes differ from the disruptive type in being specific or scenic, rather than abstract or arbitrary.

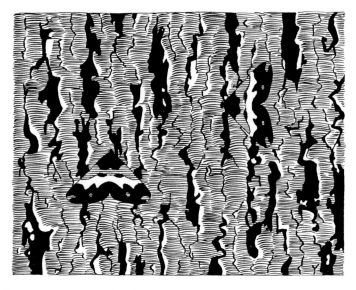

Adaptive orientation (*right*) in a moth with a transverse wing-pattern

A remarkable instance came to my notice many years ago on the Lower Zambesi, when the hawk moth known as *Xanthopan morgani* was seen at rest on the trunk of a casuarina tree. As will be seen from Plate 54, the moth attains a wonderful degree of concealment – the pattern of the fore-wings nicely reproducing that of the bark. The dark diagonal disruptive markings are well correlated with the oblique carriage of the wings, so that in the natural resting posture they form an approximately parallel series. The moth instinctively settles with its body in a vertical position, thus bringing the pattern of the wings into line with that of the bark which it simulates. Moreover, the insect applies itself flat against the surface, almost eliminating tell-tale shadow. Thus we see the interesting fact that real shadows are concealed and false shadows are suggested by the wings. Also, in any other position the insect would be more conspicuous than it actually appears in the one adopted.

In many moths a similar effect is arrived at by adaptive orientation of the body. For example, various Geometridae have the disruptive elements of the wing-pattern so disposed that when the insect is at rest, the dark shadow-markings run across at right angles to the body. When seen against a background of bark, such an insect would appear less conspicuous if it were to turn round so as to bring its body at right angles to the tree trunk. Now the interesting fact is that such moths do habitu-

Newly-hatched Crowned Plover

Squatting attitude of young Stone Curlew

ally rest in this very attitude, which once again aligns the pattern to its natural setting. The same principle is seen at work when a bittern adopts its well-known cryptic pose with beak pointing heavenwards, so as to bring the striped pattern of its throat and breast into line with the background of reeds which are its nesting place.

Deception by disguise

Many animals have taken a further step along the road to self-effacement, and obtain protection by special resemblance to some seemingly innocuous object – such as a leaf or twig, a stone or stump of wood, a piece of grass, a flower, or the dropping of a bird – that is of no interest to their enemies, or prey. In short, they make use of the principle of disguise, whereby their real identity is hidden. 'Special resemblances' occur widely in the animal kingdom and are particularly plentiful, and perfect in detail, among tropical insects. Many examples, showing the extreme modification of form and habit upon which the deceptive appearance rests, are illustrated in my book *Adaptive Coloration in Animals*. A few may be briefly mentioned here.

A small chameleon *Rhampholeon boulengeri* from the Ituri Forest, has a flattened, leaf-shaped body and a short tail that does duty as a leaf-stalk. Schmidt (69) reports of this species: 'In forest regions, where leaves of all sizes and shapes attract little attention, the strongly curved, irregular outline of the back, the dull shrivelled skin with the two peculiar dark markings suggesting the venation of a leaf, make the simulation perfect in every colour phase. At the slightest noise they usually stop in any position, even with one front and one hind leg lifted, and may remain motionless for hours.' A large toad (*Bufo superciliaris*) of the West African rain forest has the top of the head flattened and bordered with prominent brow-like ridges above the eyes which continue as folds on the flank. The upper surface is brown or grey and marked with blotches which make it wonderfully like a withered and weather-stained leaf on the forest floor. In the tree-frog *Chiromantis xerampelina* the resemblance is to bark.

Among East African tree-snakes the poisonous, back-fanged *Thelotornis kirtlandii* is one of the most specialized for arboreal life. Its body is extremely attenuated and the tail whip-like; and with its cryptic colour of variegated browns and greys it resembles a vine or liana. These snakes are sometimes in the habit of stopping with a foot or more of the body stretched forward horizontally and held motionless in space. Occasionally when in this attitude they sway the body and this adds to the

Braided Tree-frog

deceptive resemblance to a vine that is moved by a breath of wind. The only conspicuous part is the red black-tipped tongue which is used as a display organ – apparently to allure prey, which consists of lizards and small birds.

9
Advertising Coloration and Display

The possibilities of existence run so deeply into the extravagant that there is scarcely any conception too extraordinary for Nature to realise.
Agassiz

THE coloration of many animals is the reverse of cryptic, in that it serves rather to reveal than to conceal. Cryptic coloration, which tends towards effacement, is of biological value almost exclusively in the relations between predator and prey, whether used in a defensive or offensive role. Advertising, or 'phaneric' coloration, on the other hand, serves very many diverse functions – operating in the relations between rival males or members of the opposite sex, between parent and offspring, between other members of the same, or of related species, and between predators and potential prey. In this chapter we shall look at a few examples, drawn mainly from the East African scene.

Threat and courtship display

Intraspecific display is found in many groups of animals, including insects, spiders, fiddler crabs, reptiles and mammals; but it is amongst the birds that the phenomena reach a climax. Most commonly such displays serve, in one way or another, to promote successful reproduction – the ritualized postures and actions constituting a signal-language used particularly in rivalry between males, and between mates during and after courtship.

The courtship display of the Ostrich – as befits the world's largest bird – is dramatic. It begins with the male lifting and waving his wings alternately, so showing off to advantage the magnificent black and white plumes. This is accompanied by high-stepping, running and pirouetting movements: often several males join in the dance. The male then pursues the hen who, when ready to respond, sinks to the ground, the male flopping down immediately behind her to complete

Kori Bustard: at ease, and in display

his courtship. With neck lifted he now sways slowly from side to side, like a boat rolling in a heavy swell, and with each movement extends and waves alternately the wing that is uppermost. During the chase, and increasingly during mating, the whole length of the neck becomes greatly distended with blood and darkens to a bright red tint.

An important aspect of display activity is that it should be arresting: the actor is presented in an extravagant or unusual attitude. This is well illustrated by the bizarre appearance of the displaying Kori Bustard (*Ardeotis kori*). With the nuchal crest and lax feathers of the neck ruffed out, and the wings drooped, the male slowly lifts his tail, first to a vertical position in which it appears as a counterpart of the head and neck, and then forwards until its tip touches the nape, thus showing off the white, outspread undercoverts. The Kori is a bird of the short-grass plains and so is of course well known to Thomson's Gazelles who share its terrain. Yet so strange is the transformation that even they are attracted to the displaying bird. 'It is then often surrounded', writes Walther (81), 'by subadult tommies which approach with long necks from all directions. At times they flee backwards and then approach again. Finally they come so close that the big bird interrupts its display and chases them away by beating with the wings and making hissing sounds. The bustard then returns to its normal shape, and the gazelles become relaxed again.'

Species recognition and distinctiveness

Displays fulfil a variety of functions, at physiological and psychological levels – serving variously to advertise the presence of a male in defence of territory, to stimulate processes leading to ovulation, to synchronize the reproductive cycles of male and female, or to act as an emotional bond throughout the prolonged period of pair-formation, mating, nest-building, incubation and parental care.

In another category are colours and patterns which serve to differentiate closely related species. Characters that are specifically distinct tend to prevent or limit cross-breeding, and so to promote reproductive isolation. European finches, penguins of the Antarctic, and toucans of South America are good examples. The latter display patterns of yellow, red, blue and black on the beak which, in the different species, are almost heraldic, and like national flags are immediately recognized. Among African birds we see specifically distinct coloration in the wing speculum of many duck, and in the diverse colour schemes which signalize different species of plovers, turacos, hornbills, kingfishers, barbets and bee-eaters.

Anyone visiting the coastal mangrove swamps near Mombasa can observe at first hand a fascinating demonstration of this principle. Inhabitants of the mud of creeks and estuaries include legions of fiddler crabs whose burrows stud the intertidal zone, and from which the male owners emerge to defend territory against rivals, and to display to members of the opposite sex. Male fiddler crabs are characterized by the enormous development of *one* cheliped, or 'arm', which is brightly coloured and used as a display organ. Not only are the different species distinctively coloured, but each has its particular semaphore code: in one the cheliped is moved like the arm of a violinist (hence the popular name); in another it is waved on high, and seems to be beckoning; in another the crab bobs up and down as though doing 'press-ups' or 'knees-bend' exercises; in another it executes dances, circling near the burrow entrance.

Many years ago Jocelyn Crane made a detailed study of fiddler crabs of the genus *Uca* in the Panama Canal Zone. She found fifteen species, twelve actively courting, on one small beach area. 'Each species,' she writes, 'proved to have a definite, individual display, differing so markedly from that of every other species observed, that closely related species could be recognized at a distance merely by the form of the display.' These extraordinary signals govern species recognition and serve as an

42. Reticulated Giraffe, Northern Province, Kenya

Topi, Mara Masai Game Reserve

43. African Darter, Lake Naivasha, Kenya

White-backed Vulture, Mara Masai Game Reserve

isolating mechanism to ensure that mating occurs within, and not between, the several species which share the same habitat.

It is interesting to note that where related species sharing the same environment are closely similar in appearance, the distinctive signal is vocal rather than visual: such species often differ very strikingly in their calls or songs. Among the doves and pigeons, for example, peculiarities of phrase are much easier to detect in the field than differences of plumage. The following are a few examples with which the visitor to East Africa becomes familiar. To bring out the specific differences I have, where applicable, indicated syllabic quantity by morse code, accent by words, or by a verbal approximation of the uttered rhythm. Thus we have: Vinaceous Dove (*Streptopelia vinacea*): | — · · | — | , 'Cu ru cu cu,' the rhythm as in 'Crocodile egg'. Ring-necked Dove (*S. capicola*): a trisyllabic | — — — | , 'Cook her rook'. Mourning Dove (*S. decipiens*): a purring Moorhen-like 'currrroo'. Red-eyed Dove (*S. semitorquata*): | — — | · · · · | , 'My love, forget me not'. Laughing Dove (*S. senegalensis*): an unmistakable five-syllabic phrase | · · · | · · | , – this is the 'Mr O'Duffy, Mr O'Duffy' call. Namaqua Dove (*Oena capensis*): | — — | , a low-pitched 'Twooh hooo'. Speckled Pigeon (*Columba guinea*): a prolonged series of deep notes – 'Koo, who, who, who, who...' rapidly repeated. Emerald-spotted Wood-Dove (*Turtur chalcospilos*): this call is again quite distinct – a long series of notes, at first hesitant and slow, gaining in tempo and volume towards the end. The plaintive, far-away sound touches the heart, and once heard it cannot be forgotten.

The nocturnal chorus of frogs and toads is an impressive feature of the African night. Croaking, whistling, piping, trilling – every species has its own distinctive part to play in the amphibian orchestra, and the listener may detect an astonishing variety of effects, from base barking to the shrillest stridulation. As with birds, the song is of recognitional value, enabling the female to find a mate of its own species. In different African toads the call is like the bleating of sheep, the clatter of castanets, or the scraping of a comb. *Kassina* utters a single, loud, explosive whistle – 'hoip'. *Chiromantis*, the Braided Tree-frog, makes the sound of an old door moving on rusty hinges – 'crrreek, crrreek'. Little tree frogs of the genus *Hyperolius* play tinkling, xylophone notes – loud and penetrating. The Bullfrog *Rana occipitalis* – a giant among East African frogs – has a gruff, grunting chorus which I have described elsewhere as sounding like a committee of old men agreeing upon a resolution – 'um, ah, yes, um, here here, ah, yes, um, here here'.

Social signalling characters

Social recognition, or guide-marks, like those displayed in flight on the wings, tail and rump of many waders and plovers, and on the hind quarters of antelopes, serve to ensure contact between members of the flock, or to release and direct following behaviour. The patterns, disposed so as to be seen from behind, are conspicuous, simple, and specifically distinct.

East African antelopes provide interesting examples. Thomson's and Grant's Gazelle can at once be identified by the striking rear-view pattern – the white extending higher on the buttocks and being flanked by a dark stripe in the latter. The rump of Impala is again distinctive. In the Common Waterbuck, a white crescent-like ring encircles the dark rump; in the Defassa Waterbuck, the buttocks are entirely white. In his study of the social organization of this species in Uganda, C. A. Spinage noted that when the young calf is following the dam, the latter holds her tail stiffly or even vertically, and thus it remains as long as she is moving. He calls this the 'Follow me' signal. Bushbuck raise the tail, showing the conspicuous white underside, when running. Warthogs, of course, also use the tail as a signpost, holding it up like a flag-pole as they run through the grass.

In the cryptic nidifugous young of Ringed Plover, Crowned Plover and related species, a white 'nuchal' band, when displayed, assists the parent to locate and reassemble her brood. A. L. Butler, who observed Egyptian Plover in the Sudan, has described how the moment danger threatens the chicks flatten themselves in a depression. But when the young one is running about, the white patch 'can be followed by the eye to a distance at which the rest of the tiny body is invisible, looking very much like a morsel of cotton-wool being blown hither and thither over the sand.'

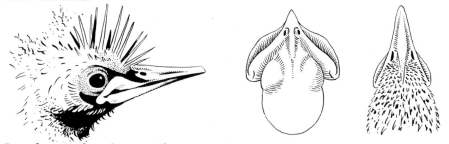

Gape-flanges of nestlings. *Left*, Hoopoe (*Upupa epops*) showing thick, conspicuous waxy-white flange at the gape-angle; *centre*, Starling (*Sturnus vulgaris*): dorsal view of head in the early nestling, showing widely-swollen mouth flanges and *right*, subsided condition of the flanges in the fledgeling (redrawn after Portman)

Feeding advertisements

The young of most nidicolous birds have the lining of the mouth brightly coloured: red, orange or yellow – the colours displayed by ornithophilous flowers – are frequent in these quasi-floral displays which, as in many flowers, often incorporate patterns of contrasting hues. For example, in the Crested Lark (*Galerida cristata*) the mouth lining is orange-yellow with three black spots arranged in a triangle on the tongue (75). The mouth of food-begging nestlings is moreover framed with greatly swollen gape-flanges which reinforce the display.

In dimly lit surroundings white is the most conspicuous colour, and it is interesting to note that the mouths of various nestlings reared in holes are adorned with pale tints rather than the saturated colours that are so effective in open situations. Thus, the fleshy flange bordering the gape of the young Splendid Glossy Starling (*Lamprocolius splendidus*) is white. The mouth of the nestling Hoopoe (*Upupa epops*) is bright pink, with waxy-white swollen gape-flanges.

More remarkable are the structures known as 'reflecting pearls' – found in nestling weavers (Ploceidae): these are beadlike bodies which 'glow like small lamps in the gloom of the covered nest and help the parents find the open mouths of young' (83).

The buccal advertisements of nestlings play a dual role: they are both 'indicators' and 'stimulators'. 'What could be more convenient', writes Armstrong, 'than that when the busy parents arrive with their dainties they should find the mouth cavity vividly painted and outlined so that they can dispose of their gleanings quickly and accurately and then rush off for more with the minimum delay' (1).

The gaping mouth of the young cuckoo has an almost compulsive effect, even inducing birds other than its own foster parents to feed it. And from America there is the record of a North Carolina Cardinal (Fringillidae) which for several days delivered food into the open mouth of goldfish in a garden pool! (83).

Distraction display

Birds of many species meet danger from intruders near the nest by actions that used to be called 'injury feigning' and are now better known as distraction display. The parent draws attention to itself, and away from the nest or young, by fluttering on the ground, lying on one side, trailing a wing, or shuffling along as if a leg was broken, and so advertising its whereabouts. Always keeping just beyond reach, the bird lures the pursuing enemy further and further away from the vulnerable nestlings.

Distraction displays are found in members of many families, including ducks, game birds, owls, plovers, sandpipers, nightjars, buntings and larks. Strangely enough the habit hardly occurs elsewhere in the animal kingdom, though something comparable is seen in the Thomson's Gazelle. When alarmed, individuals often adopt a curious gait, generally referred to as 'stotting'. With the neck erect and the head held high, and with the legs as stiff as ramrods, the gazelle bounds up and down, springing from the pastern joints. Noel Simon states that females stot when attempting to lure a predatory dog or jackal away from their fawns. Moreover, the strangeness of the gait immediately catches the eye, and so serves to alert the rest of the herd.

When stotting becomes infectious in the herd, this may confuse a pursuing enemy. Walther (81) found that adults often go off stotting from hyaenas, and regularly from Wild Dogs (but not from Lion, Leopard or Cheetah). For success the Wild Dog needs to concentrate pursuit on one individual: 'this is more difficult, the more numerous are the prey animals, the closer they keep together, and the more strikingly they run off. Because then each member of the herd stimulates the predator's pursuit.'

Warning coloration

There is an African grasshopper whose conspicuous appearance, sluggish habits and evil smell proclaim with emphasis that it is not like its fellows. Most members of the family Acridiidae are cryptic green or brown; but *Phymateus viridipes*, bright green in colour, is marked on the thorax with sealing-wax red, and has dark purple and crimson hind wings. It appears to court observation, living freely exposed to view – at Paraa it can often be seen on the oleander bushes at the hotel entrance – and scarcely resists capture. In defence it will raise its tegmina, display its red hind-wings fanwise, and dribble out from the thorax a pungent stinking frothy liquid. Many observations and experiments have shown that the insect is highly distasteful to and avoided by insect-eaters.

It happened that while *en route* for East Africa in 1941 I had been invited to give a series of lectures on animal coloration; and when during a call at Cape Town some specimens of *Phymateus* were obtained, I was able to demonstrate to my friends the effectiveness of its chemical deterrent. A keen predator was available in the person of a young bull-terrier belonging to one of the passengers. When shown the insect, the dog ran up and sniffed it, shook his head violently and began to salivate. Under further encouragement he became very excited, and made repeated though quite ineffectual attempts to get the better of it – barking, whining, warily touching or turning it over with an out-stretched paw, or quickly jumping in to make mock snaps at it. On being again urged forward, he continued to dart about, running round to approach the grasshopper from different angles, and to make pawing movements near it, as if now even afraid to touch it. Not once could he be induced to take it into his mouth. These feints continued for ten minutes; during that time the dog repeatedly shook his head as though to get rid of an unpleasant taste, while strings of mucus issued from the corners of his mouth. At the end of the experiment the *Phymateus* was unharmed and, indeed, the insect had hardly troubled to display its wings during the encounter.

Warning coloration, with associated deterrent properties, is seen among insects of many orders. Carabid beetles of the genus *Anthia* provide striking examples. Such beetles depend for protection upon

Anthia sexguttata, an aposematic carabid

their power of ejecting strongly acid liquids from behind. They are terrestrial, living in open country where their large size and bold coloration make them extraordinarily conspicuous. When alarmed they adopt a characteristic attitude, raising the body high on the legs, so that the noxious liquid can be projected upwards. Marshall (51) reported that the secretion is powerful enough to cause a strong stinging sensation when it touches the skin of the face, and that it can be projected to a distance of some four or five feet. For a night-roving creature, white is the most effective advertisement. The colour of *Anthia sexguttata* is at once striking and simple – being not unlike the six of dominoes; and is of a type easily recognized and remembered by a would-be predator.

White is also an important element in the livery of various nocturnal mammals – the skunks of America and the related weasels and zorilla of Africa. The Zorilla (*Ictonyx striatus*) is a strikingly conspicuous carnivore about the size of a polecat. Black in ground colour, the back and flanks carry a set of broad white stripes, and the tail is white. At the base of the

Honey Badger raiding a crocodile's nest

tail are the perineal glands from which it can squirt with great force the oily liquid that is its defence. If interfered with it first erects the body hairs and brandishes the bushy white brush, at the same time uttering shrill warning cries.

Another nocturnal musteline is the Ratel or Honey Badger (*Mellivora Capensis*). These animals likewise defend themselves by emitting a suffocating offensive odour from the anal glands. When attacked the Honey Badger fights desperately and is extraordinarily tenacious of life. It thick skin is like a loose coat of rubber, and Stevenson-Hamilton writes: 'I have known cases when, after a protracted struggle with a pack of dogs, the ratel has picked himself up and jogged off apparently little the worse from the encounter, leaving the assailants totally exhausted, and all more or less damaged' (74). It will even attack game, including buffalo – 'biting the groin and genital organs, the animal then bleeding to death' (21). The upper parts of the body, from crown to tail, are whitish, in contrast to the black flanks and belly. Being light above and dark beneath such a colour scheme is the reverse of countershading (see page 157) and is one well adapted to act as a warning to other animals.

A characteristic feature of aposematic animals is the seemingly nonchalant manner in which they expose themselves to view, as though courting recognition. The Marbled Tree-Frog of East Africa wears a pyjama-like suit of black, red and creamy-white stripes. The Wood-Toad *Phrynomerus bifasciata*, another East African species, is even more conspicuous, being grey and black with vivid bands of full vermilion extending from snout to groin, the legs being black with vermilion spots. Unlike most toads it is diurnal, and exposes itself to view. If alarmed or roughly handled, the toad exudes a copious sticky, poisonous, dermal secretion, which dries on the hands like gutta-percha and causes inflamation and a stinging sensation, as from nettles. The mucus kills frogs of other species that come into contact with it.

Bluff and mimicry

Warning coloration has been developed in animals of many different groups. It is associated, as we have seen, with powerful means of defence, which include such weapons as stings, poisonous flesh, distastefulness, pungent artillery and other forms of chemical warfare. Such animals are relatively immune to attack; and since enemies must learn by experience what is acceptable prey and what is to be avoided, these advertisements serve to facilitate recognition and thus favour the creatures exhibiting them.

The Marbled Tree-frog, Lower Zambesi

A different expression of the warning principle involves the sharing of a common warning livery by several unrelated distasteful species. For example, certain African beetles belonging to a number of different families, and a considerable assemblage of other insects including wasps, bugs and moths, share an orange-and-black colour pattern. Common warning coloration, or Müllerian Mimicry, provides a combined warning to enemies and tends further to promote their rapid education and thus to reduce the total number of casualties during trial-and-error stages of learning – to the mutual advantage of species in the Müllerian association.

Very different are the conspicuous advertisements and threatening attitudes displayed by certain animals which are not themselves specially armed for defence. Such animals rely upon bluff, or false warning. Their displays often involve an increase in size – actual or apparent. The Flap-necked Chameleon and the African toad *Bufo regularis* both inflate the body with air and then orientate themselves so as to present the greatest surface to an enemy. The South American 'Escuerzo' (*Ceratophrys cornuta*) distends its body to a tremendous size under provocation, at the same time uttering broken cries of remonstrance. The pug-dog-nosed African toads of the genus *Breviceps* – which appear in numbers during the rains to feed on termites – can inflate the body so as to become almost spherical.

Displays whose function is to intimidate are often sudden and, to a close observer, startling, the effect being produced by the spreading of frills or fans, by erection of hair or by mouth-gaping. The Crested or Maned Rat (*Lophiomys imhausi*), a guinea-pig-sized rodent found in forests of the Kenya highlands, adopts an intimidating pose by erecting the hair of its back and lowering that of its flank, so as to form a parting which extends the length of the body. The long hair is black at the base and white distally: so, in display, the animal's normal appearance is transformed, and there opens along each side a wide furrow, black in colour with a white border.

Among birds the sudden spreading of surface, to increase the apparent size of the displayer, is well seen in the threat posture of the Water Dikkop (*Oedicnemus vermiculatus*). This bird is always found near water and is one of the species which regularly associates with crocodiles. Many years ago Captain C. R. S. Pitman, when Game Warden in Uganda, drew attention to the regular presence of water dikkops on favoured beaches of Lake Victoria that, in those days, were 'infested' with crocodiles. On islands and banks of the Victoria Nile, and formerly on islands near Entebbe, I have frequently found eggs of the Water Dikkop on the crocodile rookeries, often within a few feet of the place where a crocodile is lying over her own eggs. Nesting seasons of bird and crocodile coincide. The birds undoubtedly serve the reptiles well as watch-dogs. It is also possible that from the habit of nesting beside the guardian crocodiles they may gain some adventitious protection against egg-eating enemies. But they also have their own effective threat display. If approached by a marauding Nile Monitor, the sitting bird will run from the nest to confront its enemy, its wings widely spread and rotated so that the vertical fan of feathers makes its full impression: at the same time the tail is cocked and fanned out. I have seen these exhibitions of threat when watching crocodiles from a hide and there is no doubt that they act as a deterrent to the relatively formidable intruder – an example of psychological warfare. The monitors are also harried by Pied Kingfishers that have their nesting holes in steep banks behind some of the crocodile grounds. Sometimes four or more of the kingfishers will mob and repeatedly dive-bomb the monitor, and during one such attack the lizard retreated at speed, bleeding, after a kingfisher's beak had found its target just behind the eye.

Organs exposed or erected in display sometimes carry a large ocellus, or dummy eye. A beautiful example is the East African mantis *Pseudocreobotra wahlbergi*. When alarmed the insect raises its wing-covers over its back like two signal arms – each of which bears a conspicuous ocellus – thus directing towards its enemy a message of intimidation.

Many moths carry on the hind wings a pair of staring eye-spots. In some, the spots are permanently exposed; in others they are concealed by the folded-back fore wings when the insect is at rest. If discovered, the moth has a second line of defence in the sudden and startling exposure

Warning display of the mantis *Pseudocreobotra wahlbergii*

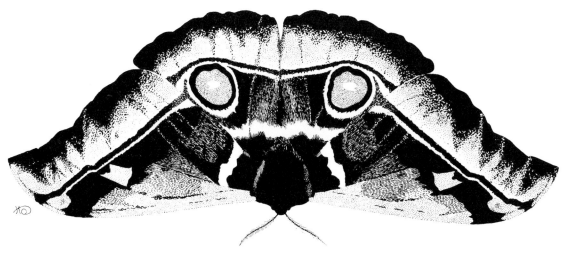

Bunaea alcinoe displaying eye spots

of two eyes. A familiar example in the Sphingidae is the Eyed Hawk-moth. This mode of defence is highly developed in many African Saturniidae. *Bunaea alcinoë* is one I came across at Lake Manyara and is here shown in its display attitude. Some species have even more realistic 'frightening eyes', with a large black pupil, flaming iris, and light reflected from the cornea all simulated. The ocellus of *Lobobunaea phaedusa* is 'an exact copy of the eye of a wild cat'. In case of attack the effect is enhanced by the resemblance of the raised fore wings to the ears of a mammal; and at close range it must afford the most terrific sight to birds and other insect-eaters that are themselves the favourite food of such carnivores as genet, serval or civet.

That these 'terrifying eyes' are an effective defence against insectivor-ous birds has been demonstrated in ingenious experiments carried out by A. D. Blest (*Behaviour*, 1957). Reactions of passerine birds – chaffinches, buntings, tits and starlings – to a butterfly's threat display were tested by suddenly projecting the eye-spot pattern of a Nymphalid butterfly immediately alongside mealworms that the birds were about to eat. Appearance of the projected patterns released the escape response of the birds. Moreover, birds that had been hand-reared in isolation from their own predators and which had never before seen similar patterns nevertheless showed full escape response to the simulated displays. Conversely, as reported by Niko Tinbergen, while experiments showed that birds were scared away by the threatening display of the Eyed Hawk-moth, 'When the colours of the hind wings were brushed off

. . . the display did not make the least impression on the birds, and the hapless moth was eaten forthwith.'

Colour conflict

We have seen in the previous and present chapters that the phenomena of adaptive coloration fall broadly into two main functional categories: cryptic – those that conceal; and sematic – those that reveal. Both concealment and advertisement carry biological advantages. For example, conspicuousness is essential for display: yet this runs counter to the need for concealment. The two types of coloration are antagonistic, if not mutually exclusive. There is thus a conflict between the rival claims of cryptic and sematic coloration, and it is interesting to note how this difficulty has been met. In the short account which follows we may take examples mainly from African birds.

Firstly, the conspicuous phase may be transitory. In many insects and birds where the coloration is predominantly cryptic in the resting attitude, previously-hidden advertising characters are suddenly exposed to view in movement – especially during flight, or in display attitudes. Such are the group-recognition marks on secondary wing-feathers, rump and tail of many plovers, waders, duck and chats; and the feeding releasers of nestling birds. An interesting example is the Black-throated Honeyguide – the species that guides man and the Ratel to bees' nests. When guiding the bird attracts attention to itself by incessant chattering and by showing off conspicuously the white outer tail feathers.

A dual-purpose dress is seen in animals which exhibit what have been called 'flash colours' when they leap or fly. Many tropical grasshoppers which are extremely well concealed when at rest, have hind wings of red or blue, yellow or purple, often outlined with a black border, which renders them at once conspicuous in flight. But on landing, the hind wings are instantly hidden and the contrast is so sudden that the insect seems to vanish. Similarly some tree-frogs have brilliant colours on parts of the body and legs which are hidden until the frog leaps.

Again, the problem of colour conflict may be met by the use of two rival sets of coloration in different groups of individuals of a species. In such cases the advertising dress is worn by the less vulnerable, and the cryptic dress by the more defenceless individuals. Thus, in sexually dimorphic game birds and duck which nest in the open, selection for conspicuousness can have full play in the male, while the female has been forced to take the evolutionary road to concealment. Parallel conditions obtain in the parent-child relationship: for instance the

44. Nile Monitor eating crocodile's egg, Kabalega

Newly-hatched Nile Crocodiles, Victoria Nile

45. Nile Crocodile mouth-gaping, below Kabalega Falls

46. Flap-necked Chameleon, Paraa, Uganda

Male Agamid Lizard, Serengeti National Park

young of many ground nesting species such as plovers, Avocet, African Jacana, Skimmer, and of gulls and terns, in which adults of both sexes are conspicuous, are themselves highly cryptic.

In the case of birds that are relatively non-vulnerable because of large size, fighting strength, or gregarious habits, the path towards conspicuousness lies open and has been taken by both male and female for use in display and social life. Notable examples in East Africa include pelicans, flamingoes, cormorants, storks, egrets, various large birds of prey, Ground Hornbill and White-naped Raven. The same applies to birds living in dense forests – such as turacos, parrots, the larger hornbills, some barbets, orioles and others – where the need for visual concealment is least. On the other hand, among ground living or otherwise defenceless birds of open country, the advantages of concealment override those of advertisement, and bright display colours are generally absent: quails, coursers, stone-curlews, nightjars, sand-plovers, francolins, grass-warblers and larks are cryptically coloured in both sexes.

There are many exceptions among small birds living in open country which are nevertheless highly conspicuous. My attention was first drawn to a possible explanation of this anomaly when in 1941, during a week's leave in Middle Egypt, I happened to be preparing some bird skins. The carcass of a Palm Dove and of a Pied Kingfisher had been discarded; and it was soon noticed that hornets – which are omnivorous scavengers – were attacking one carcass only, that of the dove; the kingfisher they neglected, evidently finding the flesh distasteful. This chance observation led to a series of experiments in which hornets were used as tasters to assess the relative palatability of birds' flesh (9). Of thirty-eight species the three found to be least edible were the Pied Kingfisher (*Ceryle rudis*), White-rumped Black Chat (*Oenanthe leucopyga*) and Mourning Chat (*O. lugens*) – all highly conspicuous birds with entirely black and white plumage. In contrast the two heading the list as most palatable were Crested Lark (*Galerida cristata*) and Wryneck (*Jynx torquilla*), both highly cryptic species.

Sixteen years later during a research visit to Zambia, I was able to extend the investigation. With the kind and enthusiastic help of Mr C. W. Benson an extensive series of palatability tests was carried out by a panel of tasters recruited from staff of the Department of Game and Tsetse Control, Chilanga. Again in this series, comprising nearly two hundred species, those rated as most palatable were predominantly cryptic birds such as White-backed Duck (*Thalassornis leuconotus*), Quail (*Coturnix coturnix*), African Crake (*Crex egregia*), Senegal Bustard (*Eupodotis cafra*), Water Dikkop (*Burhinus vermiculatus*), Double-banded (*Pterocles bicinctus*) and Yellow-throated Sandgrouse (*P. gutturalis*) and Fiery-necked Nightjar (*Caprimulgus pectoralis*). At the other extreme, species with the lowest palatability-scores contained a high proportion of conspicuous birds. Small species with entirely black, or black and white plumage – including Pied Kingfisher, Black Crake, Black Cuckoo (*Cuculus cafer*), White-headed Black Chat (*Myrmecocichla arnoti*), Black Chat (*M. nigra*), White-winged Black Tit (*Parus leucomelas*) and Black Tit (*P. niger*) – were all found to be bitter, sour or otherwise markedly distasteful (15).

Among these small and otherwise defenceless species, there is thus an inverse relationship between acceptability of the flesh and visibility of the plumage. Such birds illustrate yet another means by which the conflicting needs of effacement and advertisement may be reconciled. In many, as we have seen, palatability to enemies tends to render cryptic coloration of paramount importance. Another line of adaptation has led in an opposite direction, towards conspicuousness, a trend that has become possible when it is associated with the counter-deterrent of repugnant taste.

IO
Animal Portrayal

How small is the power of words to convey clear notions of visible
things and on the contrary how well fitted for this task is the craft
of the limner.   *Leonardo da Vinci*

THOUSANDS of years ago animals were the object of man's earliest
artistic endeavours – surviving today in cave drawings from the Ice Age
and reliefs of lion hunts in ancient Mesopotamia. Whatever may have
been the motive, practical, magical or aesthetic, that inspired these
murals and carvings, the lively hunting scenes reflect not only high
artistic skill but also the delight with which primitive people approached
their craft.

Today tens of thousands make a pilgrimage each year to the game
areas of East Africa to seek enchantment in 'the land of wild beasts', and
in their turn make a pictorial record of unusual experiences and fascinating encounters in the wilderness.

A word about cameras

It may almost be taken for granted that the modern traveller will carry
a 35 mm camera loaded with colour film. The present day popularity of
the miniature format serves to demonstrate the extraordinary transformation of photographic equipment that has occurred during the
century. The pioneering photographers of bird life and big game used
cameras that by modern standards were enormously cumbersome; and
of course glass plates rather than roll film were the order of the day.

But it must not be supposed that the tourist with his sophisticated,
expensive, miniaturized and automated gear will necessarily produce
better pictures. Indeed, there is no direct connexion between the
quality of a photographer's work and the cost and complexity of his
equipment. Astonishing photographs, which we may regard with
feelings of wonder and humility, were obtained early in the century by
men like Cherry Kearton, Herbert Ponting and R. B. Lodge – using
simple and inexpensive apparatus.

'What kind of camera do you use?' is the question one is so often asked. The choice of photographic equipment for an expedition, or safari, is not easily made. There is no such thing as the ideal camera. Selection will depend upon the way the camera is to be used in the field, and the photographer's aim – for example, to make exhibition prints in black-and-white, or records for research, or colour transparencies for projection or reproduction.

My own early experiments in tropical photography began just fifty years ago, in Brazil. At that time I used folding plate cameras with ample bellows giving 'double-extension' which made possible the photography of small objects at life size. One of these was a Phoenix quarter-plate and another a Contessa Nettel taking 10 by 15 cm plates (postcard size). For photography of birds, insects and other animals I continued to use quarter-plate cameras (even preferring glass plates to film pack) for thirty years, having added a very solid teak-and-brass Sanderson tropical model to the collection. When mounted on a tripod these instruments were excellent for use from the hide and gave satisfactory service up to 1952 when I was beginning studies of the Nile Crocodile in Uganda. Subsequently I went over to films, for greater convenience when travelling – first using a twin-lens Rollei which was later replaced with a Telerolleiflex, taking pictures 6 cm square. Only in recent years have I taken to the 35 mm format, mainly for colour work when on safari.

Quite apart from the fact that it is so handy for rapid spontaneous snapshotting (I do not use the word in any derogatory sense), the miniature camera has several other important advantages over the larger instrument. One is that the makers supply a standard (50 or 55 mm) lens of very wide aperture: the greater speed associated with a wide f/stop greatly increases the scope of the 35 mm – for example, when fast exposures are required in poor light or in artificial light. An even more important attribute of the standard lens of short focal length is its increased depth of focus: it is here that the 35 mm outstrips the performance of larger cameras. The speed with which films can be changed is now taken for granted. In a minute the camera is made ready for 36 new exposures. But it was not always like this. I am reminded of former days in camp, when plate-holders had to be reloaded by feel and when, with the arms encased in a changing-bag, one suffered from sweat, frustration and tsetse flies.

The claim is sometimes made for the miniature camera that it gives as good an enlargement as the larger field instrument. If this were true, then one may well wonder why it is that the precision 35 mm has not replaced the more bulky cameras used in many special fields – such as group portraiture, aerial survey, archaeology and astronomy – where negatives containing a maximum of detailed information are required. When I compare 10 by 12 inch enlargements made from my quarter-plates – an example is the hawk-moth taken in 1927 – with those made recently from black-and-white 35 mm film, I am in no doubt that the smaller negative just lacks the required critically sharp detail and quality of the larger negative.

It is also worth remembering that the best arrangement, as composition, will hardly ever be seen in a print made from the *whole* negative. Here again, the larger format scores – even if it is no larger than the square 6 by 6 cm – in the latitude it allows for planning the picture in the enlarger. With it we can arrange the proportions and orientation of

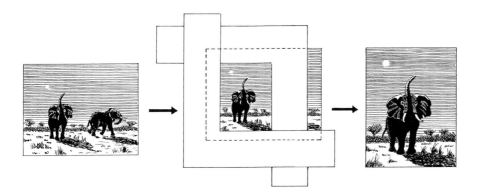

Left, print from whole negative; *centre*, the print masked; *right*, enlargement from part of the negative

the rectangle, exclude unwanted areas of foreground or background, and place the principal subject where we want it in the frame. Sometimes in so doing we may use only a small part of the negative: examples of such treatment will be found in my book on zoological photography (12).

Lenses and accessories

The choice of film to be taken on safari – whether black-and-white or colour, fast or slow, this or that make – is a personal one: it will depend upon the kind of work to be attempted. Many experienced photographers use black-and-white film. This medium has the distinct advantage that everything is under the user's control, from the initial exposure, through development of the negative, to the finished enlargement.

Where the colour film excels is in its unique quality for projection. Greatly magnified on the screen a first-class colour transparency creates a wonderful illusion of reality that can never be matched in the colour print or book illustration. The difference is due to the conditions of illumination: the print is seen in reflected light, and its brightness scale (from light to dark) is limited; whereas the brightness scale of the transmitted picture depends upon the intensity of the light source in the projector and upon the stopping power of the deepest shadows in the slide – it is virtually unlimited and can give effects of astonishing luminosity and brilliance.

A further consideration is film-speed. The manufacturers provide a wide range of choice. It should be emphasized that the slow film – compatible with demands of the subject – is to be preferred. Slow emulsions give negatives or transparencies of superior quality both as regards tonal gradation, clear definition and grain structure. This is another way of saying that 'the only advantage of high speed film is high speed'. For colour transparencies many professional photographers stick to Kodachrome II, which at ASA 25 is perhaps the slowest colour film on the market: it is certainly one of the very best to take on safari.

The keen photographer of wild animals will probably be using a reflex camera with interchangeable lenses. And here I hope a word of advice may be acceptable to the less experienced camera user. Distant objects appear surprisingly smaller in the photograph than they did to the observer at the time the photograph was taken. Even when one is close to game the photograph taken with a standard 55 mm lens can be very disappointing. It must be remembered that the linear size of the image on the film is directly proportional to the focal length of the lens. The elephant that at fifty yards distance appears in the mind's eye as a huge object eleven feet tall is accepted by the camera as a tiny image some 8 mm in height and looking disproportionately small in the 36 by 25 mm frame.

A 'telephoto' or long focus lens is an essential item to be taken into the field. Some people prefer the versatile zoom lens with its variable focal length. In my experience 135 mm is most acceptable for general use, though when one is photographing small animals or unapproachable game a 300 mm lens may be required – as it certainly will be if birds are the target. For special subjects needing a 600 mm lens the camera will generally be mounted on a tripod. It should also be borne in mind that the telephoto lens may be used to give improved perspective in

Illustrating changes in perspective that come with distance. *Above*, the subject as recorded with a standard lens; *centre*, the same subject seen from further away; *below*, the centre picture recorded with a lens of longer focal length

composing the picture, and as explained in the figure above, even to move mountains.

Finally, a brief word about filters. Landscapes, especially at high altitudes, reflect excessive ultra-violet light (not visible to the eye but recorded on the film) which may give the landscape an unnatural bluish tint, or fog the black-and-white negative. An ultra-violet filter cuts out these unwanted rays. U.V. filters do not materially alter the exposure time; and since they also serve to protect the lens from mechanical injury they should be permanently fitted.

The polarizing filter has a more specialized role. It is selective not for wave-length but for the plane of vibration of light rays. The filtering effect alters as the filter is rotated in front of the lens. Suitably adjusted, the polarizing filter cuts out light reflected from smooth surfaces such as water, thus suppressing excessive glare and greatly improving the range of tones in the composition. Under appropriate conditions these filters can enormously enhance the rendering of clouds and distant landscape.

Working from a hide

There are two different methods of approach to wild life photography. One is planned and deliberate work from the hide; the other opportunist and spontaneous snap-shooting from a vehicle, or on foot. For the first, the camera will be mounted; for the second, hand-held. In the one the animal comes to the camera; in the other, the photographer seeks the animal. Siting of the hide demands knowledge of the animal's habits. The chosen site will be one offering some natural attraction to which the quarry will come to feed, or drink, or shelter – a water-hole or salt-lick, a burrow or a basking place, a nest or a carcass. For hide work anywhere patience is an asset. On the other hand, for the open approach, quick decisions, practised handling of the camera, and luck all contribute to success.

In a country where natives are expert at fieldcraft, there is no need to travel encumbered with a prefabricated hide. With local labour and local material a hide can be erected at short notice. It will be made of grass thatch supported on a framework of sticks – a construction that allows of free ventilation, so important in a climate where the temperature inside a fabric enclosure would soon become intolerable.

With long experience of waiting and watching in hides overlooking crocodile rookeries I have found that it is a mistake to economize in space. The hide should have a grass roof for shade, and should be high enough to allow one to stand erect and roomy enough for free movement. It will have a camouflaged window not only facing the 'stage', but one on each side, since the photographer must be prepared for visitors such as hippo or elephant that may approach from any direction. It is useful to have knowledge of a climbable tree nearby.

Working from a hide, as compared with spontaneous snap-shooting from a launch or Land-Rover, has many advantages. The camera may be larger than miniature-size and it will be secure on a tripod. There is

47. A large bull Crocodile defending its river-bank territory below the Kabalega Falls

A female Crocodile, lying over its nest at Fajao, guards its buried clutch from marauding monitors and baboons

48. Crocodile eggs hatching: at birth the babies measure about thirteen inches in
length

Maternal care: the female guards hatchlings in the nursery pool. Two babies ride
on her head, Paraa

49. Lesser Kuda and Impala, Tsavo West

Oribi at Buligi in Kabalega Falls Park
Both photographs illustrate the illusion of flatness produced by counter-shading

50. Leopard surprised in long grass and ready to attack, Serengeti National Park

51. Long-tailed Nightjar (*Scotornis climacurus*) photographed by flashlight, Paraa, Uganda

Water Dikkop (*Burhinus vermiculatus*). This bird commonly nests beside nesting crocodiles in Uganda

52. Jackson's, or Three-horned Chameleon. Coloration and form combine to make the chameleon inconspicuous on a lichen-covered twig. Nairobi

A frog, *Rana oxyrhynchus*, concealed in its natural surroundings by disruptive coloration, Jinja

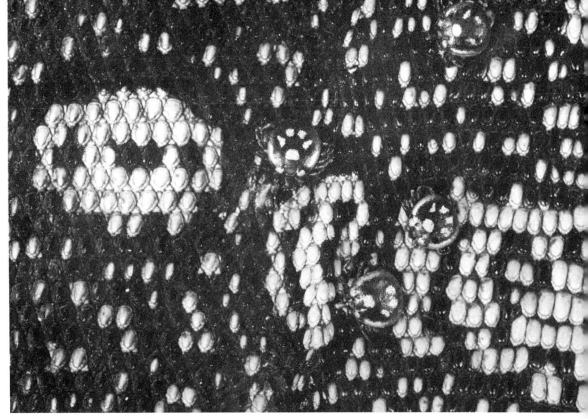

53. The lizard tick *Aponomma exornatum* in situ on the scales of a Nile Monitor. Coloured dark brown with yellowish spots, the ticks harmonize closely with the colour of their host

Flashlight photograph of the gecko *Pachydactylus bibronii* seen against the rock which it resembles in colour, Kafue River, Zambia

54. The Hawk-moth *Xanthopan morgani* in its natural attitude of rest on a Casuarina tree trunk, Beira: a remarkable example of background picturing

55. A savanna Grasshopper, *Certacanthacris tatarica*, wears a disruptive disguise of dark brown, pale buff and grass-green, which effectively conceals it in its natural surroundings. Nairobi

An aposematic Grasshopper, *Phymateus viridipes*, of sluggish habit, conspicuous appearance and evil odour, Mortimer, Cape Province

56. Mother and child, Amboseli

57. Elephants browsing in ground-water forest, Lake Manyara

Bull Elephant feeding on papyrus, Lake Amin

58. Masai Giraffe beneath its food plant, *Acacia tortilis*, Manyara

Giraffes feeding on thorn acacia, Seronera

59. Portrait of a Lion, Serengeti National Park

60. Buffalo at Fajao, Uganda

61. Afternoon at Treetops: the Waterbuck come down to drink

Waterbuck and aloes, Seronera

62. Study of a spotted cat – the Serval

plenty of time for camera adjusting and picture planning. Being concealed from his quarry the photographer can watch and record his subjects behaving naturally and at close range. Too often pictures taken in the open show animals looking suspicious, alerted to danger, or in the act of moving away.

The photographer on safari

Photography from the hide is mainly for the specialist and research worker. Most visitors to the game areas will use the hand-held 35 mm camera, and will be taking photographs from a launch or motor vehicle. I shall deal here with a few points that become important for such opportunist shooting.

And first, a word about the need for constant preparedness. Opportunities come and go very quickly. One may suddenly find a dik-dik standing by the roadside, or an elephant and calf may come into view beautifully framed between forest trees. That is not the time to discover that your film is finished. When only one frame of a film is left, it is wise to reload and be ready for whatever thrilling action may lie ahead, rather than to be caught unawares.

The best time for wild life photography in the tropics is during the first hours of daylight and again in late afternoon and evening. There are two reasons for this. Firstly, in the cool of morning and evening mammals and birds are active and are more likely to be encountered where they can be seen to advantage. In the heat of the day many birds become silent and take to cover, and game animals will be resting and sheltering in shade. Secondly, there is the question of lighting. As I have mentioned elsewhere (12), a high or near-vertical sun is as unsatisfactory in the field as a single overhead lamp would be in the studio. No photographer would willingly work under these conditions. They deny him any choice between flood or side illumination or *contrejour* effects, such as low oblique lighting affords. A strong top light is always undesirable pictorially, whether in animal portraiture or habitat studies: under a high sun the animal stands or sits above its short, dark shadow; trees become symmetrical in form, and trunk and branches are lost in heavy shade. Shadows themselves are dense and lacking in detail, because less scattered light can reach them from the sky when they lie beneath, rather than beyond, the object that casts them. And lastly, heat haze has to be reckoned with; for in the heat of a high sun distant objects near the ground quiver and are subject to the distortions of mirage.

People who have not travelled near the equator often imagine that the light there is likely to be ideal for photography. How often the photographer, home from the tropics, will hear the remark: 'Of course, the light must be wonderful.' The speaker implies that once you have left behind the grey skies of northern latitudes, you can scarcely fail to obtain striking results. He imagines a scene where the sun is always hot, the light always bright, and where short exposures and small apertures are the order of the day. Hot it may be, but heat rays and actinic light are different things. And the quality and brightness of the day can be deceptive.

Many modern 35 mm cameras are fitted with rapid-reading through-the-lens exposure systems, and needles in the view finder indicate the exposure suited to any combination of shutter speed and lens stop. But such readings are not infallible. For example, when taking a dark animal such as a buffalo against bright light reflected from the land-scape, the meter reading will give a correct exposure for the sky, rather than for the subject. The old advice is still sound – that we should expose for the shadows and let the highlights take care of themselves.

When photographing fast moving subjects which require short exposures of 1/500 or 1/1000 of a second, or subjects that are dimly lit, high speed films such as Ilford HPS or Kodak Tri-X Pan are required. The latter film is rated at 400 ASA, but by suitable development it can be forced up to a rating of 800 or even 1200 ASA. In Kenya the famous 'Treetops' at Nyeri, Mount Kenya Lodge near Nanyuki, Samburu Game Lodge in the Northern Frontier and Kilaguni Lodge in the Tsavo National Park all have 'artificial moons' to flood-light the stage where game come after dark to salt-licks and water. Photography of these animals is well within the reach of the keen amateur who is equipped with high speed film. The correct camera settings will of course depend upon the nearness of the game to the light source. The following table gives, as a rough guide, suggested exposures for 'Treetops' photography when animals are at about fifty yards range.

F Stop	Film speed ASA			
	100	200	400	1000
1·4	1/8	1/15	1/30	1/125
2·8	1/2	1/4	1/8	1/30
5·6	2	1	1/2	1/8

The ease with which the modern 35 mm camera can be used should enable the user to concentrate his attention on the subject. But all too often in the excitement of the moment quick snapshots are taken with insufficient care or thought for the arrangement of the subject in the picture frame, or even for such technicalities as correct exposing, accurate focusing and steady handling. Everyone knows how easily pictures can be spoilt through last-second mistakes.

Difficulties are increased when we are using a telephoto lens, both because the depth of field is reduced and because movement of the camera is magnified. Thus while care in focusing and steady handling are important at all times, they merit special consideration if we are using a 200 mm or 300 mm lens.

To secure absolute sharpness – the best the lens can give – it is helpful to focus on some part of the subject having clearly-defined and conspicuous detail: examples are the tusk of an elephant, the whiskers of a lion, the teeth of a crocodile, or the eye of a bird. Sometimes a good target for critical focusing is provided by an adjacent twig or grass-blade. When photographing small subjects at close range with the lens almost fully extended, it is best to focus *not* with the lens cylinder, but by bringing the camera into the precise taking position.

It must be remembered that loss of sharp definition is often caused by slight movement of the camera during exposure. There are several methods of reducing camera shake. Firstly, the exposure should be short – not more than 1/200 second when a telephoto lens is mounted. If one is standing in the open the camera should be held in such a way that its body and the thumb-knuckles are pressed tightly against the face, so that the camera can only move with the head. A telescopic monopod stand can be useful: while not a rigid support, it does eliminate pitching and rolling movements of the camera.

Owing to vibration it is impossible to hold a camera steady in a launch or minibus when the engine is running: so the first step is to switch off. Some photographers use a gadget which will clamp on to the door of a car, but the method is cumbersome and the camera is unlikely to be at eye level, which is where you want it. With the window raised to a convenient height one can brace the camera against the door frame: and it is here that a sock or small bag, filled with rice, can be invaluable. With the rice-bag draped over the window, and the barrel of the lens nestled into a groove, steadiness is perfect – if only other passengers will keep still for the shot. Finally, the shutter release must be depressed as carefully as you would squeeze the trigger of a rifle: this requires concentration and practice. A quick snatch at the release is fatal.

Development in the field

During expeditions I have always developed black-and-white negatives in the field. Once they have been exposed, negatives do tend to deteriorate in a hot climate, and the sooner processing can be done the better. It is also useful to see one's results at once while there may still be an opportunity to replace a failure.

In the early days, when one used orthochromatic plates, it was convenient to develop by inspection. Often the work would be done in remote spots, and I travelled with an old-fashioned safe-light: this museum piece was known as the 'Folding Ruby Fabric Dark Room Lamp' of Kodak. It was compact, and its light source a child's night light. With panchromatic material one had to heed the warning on the packet: 'develop in total darkness'. It is remarkable that with practice the manipulation of plate-holders and the handling of plates or films in the dish is done as easily by touch as by sight. But there remains the problem of finding a dark-room. Looking back over the years I recall many bizarre quarters that were turned over to photographic use, often under strange circumstances.

In my experience lodge and hotel managers were generally most tolerant and helpful; and from time to time I have been given access to linen-cupboards, larders, store-rooms, cellars and assorted out-buildings. During weeks of field work at Paraa, before the Lodge was built, I was allowed to take over a roomy corrugated-iron latrine in the tented camp. This was converted into an abode dark enough for use at night, and a notice was fixed to the door 'Strictly private. Staff only'. It was in this place that one was liable to be temporarily trapped in the early hours by roaming elephants.

When one is in camp, a tent will on a moonless night provide all the darkness needed. Even so, interruptions may occur. One night in 1952 when I was developing a batch of quarter-plates in open dishes, rain began to patter on the canvas: soon the downpour became a deluge. The wind roared. The storm that broke was terrific, and the lightning a wonder to behold. I was as much impressed with the celestial pyrotechnics as disappointed at having some plates fogged. The photographic session had to be abandoned; and I jotted down this triolet in my journal:

> The lamps of God shone out tonight,
>   When lightning cleft the Nile's dark road
> With searing shafts that blinded sight.
> The lamps of God shone out tonight;
> While firefly beacons, flashing bright,
>   The verge with golden star-dust strewed.
> The lamps of God shone out tonight
>   When lightning cleft the Nile's dark road.

Storm at night, Victoria Nile

197

I have attempted to recall the scene in the drawing on page 197 which was done months later at home.

The fact is that no one wishing to see his results in the field need be discouraged by lack of facilities, if only he is prepared to keep late hours. Then the world is our dark room. Perhaps I should add that nearly all the monochrome photographs reproduced in this book were developed wherever I happened to be working or travelling, and with a minimum of equipment – a timing-clock and thermometer, a few dishes for developer, stop-bath, and hardening-fixing solution, and two or three buckets of washing water.

Drawing with the pen

Lastly, in making a few remarks about the pen illustrations in this book I hasten at once to say that I am well aware of their various short-comings. When a drawing does not come off as we wish, at least we have learned something: we discard it and try anew.

Anyone can draw, after a fashion. And even though we may not excel, the attempt is rewarding. For drawing provides a wonderful distraction from the day-to-day uncertainties of life. While we are planning, carrying out and completing the illustration – whether a rapid impression or a more elaborate study – the mind is entirely absorbed with the work in hand. Moreover, drawing helps us to observe. Frederick Franck in *The Zen of Seeing* says: 'What I have not drawn, I have never really seen'. Through the act of drawing we increase awareness, take renewed delight in nature, and see a familiar thing with surprise as though for the first time. And if we try to represent it as if it was the most precious thing upon the earth, we shall learn to approach nature with humility.

I hope that perhaps some who read these lines and who, like the writer, have never attended an art school, will take courage to add a sketch book to their safari kit. Drawing is indeed to be recommended as a wonderful pastime. By experiment, by study, and by constant practice we gain in confidence; and in time develop a sure touch and a style as individual as is our own hand-writing.

As a means of expression and communication the pen is a splendid instrument – if only one can find out how to use it. And here one calls to mind Ruskin's advice to the student: 'Do not therefore torment yourself because you cannot do as well as you would like; but work patiently . . . The best answerer of questions is perseverence; and the best drawing masters are the woods and hills.'

The pen sketch or drawing has a quality of finality. It is crisp and

Crocodile crossing a sand river

Above and *right*, Crocodiles entering water

clear-cut as a stencil. It is also a nearly perfect medium for book illustration, being the only one that will give an exact image of the artist's work: for the reproduction is not a translation in half-tone, but a replica line for line and mark for mark of the original work.

A distinguishing feature of the pen and ink drawing is its capacity for intense contrasts of tone. The blackness of the Indian ink stands out from the whiteness of the paper, and if properly managed the pen work has a bright and almost sparkling quality. Several of the drawings in this

Burchell's Zebra grazing

Baboon with crocodile's egg

book consist simply of black patches on the paper. The viewer is given no clue from effects of light and shade or from outline. But the power of suggestion enables him to see what is missing – that is to say, the surfaces of the subject and their boundaries. This use of ink is seen, for example, in the drawings of cheetahs, serval, vultures and zebra. In the figure on page 159 blots of ink – variously shaped and arranged – represent the dark shell-markings and the dark interspaces of ground vegetation. It is left to the observer to interpret the pattern and to see the four eggs in the nest.

A different use of black masses is seen in the figure on page 137: the crocodile has just hauled out of the river; the scales are wet and gleaming. Seen against the glare of fiery sky and sun-baked earth, the crocodile and the canopy of leaves stand out strongly. The various surfaces are drawn in solid black where they lie in shade: the white paper is left immaculate, to suggest the heat and radiance beyond.

The pen is perhaps a better instrument for suggesting texture than any other art medium. A few black marks, rightly disposed on the paper – and the magic of illusion is invoked. Several of my drawings – such as they are – illustrate textural renderings of skin and scale, fur, feather and chitin and of background objects such as wood, mud and the like; and a brief reference here to different effects produced by the pen may not be out of place.

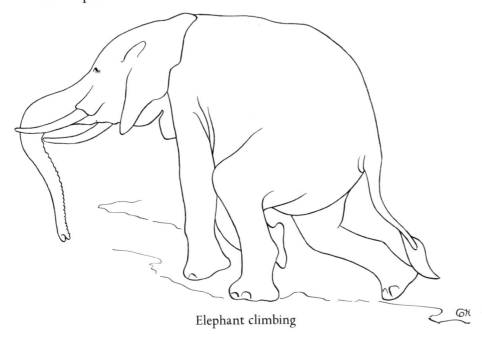

Elephant climbing

Necking display

The drawing on page 180 of the delightful little tree-frog called *Hyperolius marmoratus* was based on a photograph which I took forty-seven years ago during a collecting expedition in Portuguese East Africa. The conspicuous warning livery of the frog is here rendered in stripes of black, white and grey: the texture of the skin is suggested with a stippled tone where it lies in shade; the highlight on the black eye reveals the cornea as a polished, reflecting surface; parallel and somewhat broken lines of the maize plant represent the striated texture of the stem, which is marked by the work of leaf-boring insects; and the nondescript background sets off the frog and its perch by tonal contrasts. This interpretation in black-on-white seems to me – if I may say so – far preferable to the half-tones of the original photograph.

A different problem in textural rendering was posed by the gigantic dung-beetle *Heliocopris dilloni*, page 16, which flew on to the terrace of Kilaguni Lodge one evening while we were dining, and which was later drawn direct. Here an attempt has been made to suggest various surface qualities of the insect's chitinous armour. The elytra are smooth, polished and finely-pitted, with shallow longitudinal grooves. The thorax has a raised saddle-like eminence, rugose in texture. The head – used like a shovel – is wedge-shaped, concave in the median plane, and rises to a sharp transverse ridge: it has the texture of morocco. The legs are burnished, and their form is complex. This is one of the world's most powerful insects, and the drawing is intended to suggest strength of body and limb.

Markedly different in treatment is the drawing of a Saturniid moth, page 183, where stipple is used to suggest the soft and dusty quality of a moth's wings. In the Jackson's Chameleon, page 20, pen work is intended variously to represent the granular, pleated skin, bead-like scales, and the ornamental armoured head with its swivel-eye. The chameleon rests on lichen-covered wood, and is offset against a leafy background with interchanged tones.

The Eagle Owl, page 132, is portrayed in threat display. Here again an attempt has been made to reveal different surface qualities: the smoothly-sculptured dead timber from which the bark has fallen away; the fluffy breast feathers, contrasted with the strong raptorial claws; the foreshortened feathers of the throat which are held erect to look like a dark moustache; and the large gleaming eyes which show menace.

The question is sometimes put: 'How long did it take you to do this?' The speaker seems to consider each drawing in terms of so much toil in time. It is true that simplicity and economy of effort are often to be recommended. The rough sketch of a baboon contemplating its breakfast, page 202, was quickly done. So was the climbing elephant, page 203, which happens to be the only outline study in the book. But to produce

Cheetah stretching

Hugh B. Cott.

Aristocrat of the Felidae

other effects much detailed drawing may be required. When at work the artist does not watch the clock, and indeed the passage of time is almost unnoticed. Alfred East, the landscape painter, said: 'Do not go to your

Bull Elephant, Ishasha, Uganda

work as a task, but as a labour of love.' And so it is: the real work – a sort of hidden asset – that goes into the drawing comes not from the hand, but from the heart.

Polychrus acutirostris: a convergent Brazilian counterpart of an Old World chameleon

As we traverse the miles of African wilderness – by lake shore, river and saline swamp, through lush gallery forests or alpine moorland, over limitless short-grass plains, acacia parkland and arid thorn bush – scenes we wish to memorize and record present themselves in superabundance. The eye of the photographer and artist will for ever be on the look-out for the beautiful. But where is that to be found? A painter recently remarked: 'Beauty is everywhere, except where man has been.'

208

Journeying through the unspoilt wild lands which still appear virtually as they were ages before man appeared on the scene, we cannot fail to recognize the truth in this generalization. But we may also recall the words of Emerson: 'Though we travel the world over to find the beautiful, we must carry it with us, or we find it not.'

References

1. Armstrong, Edward A. (1947). *Bird Display and Behaviour*. London (Lindsay Drummond).
2. Bannerman, David A. (1930–48). *The Birds of Tropical West Africa*. London (Crown Agents for the Colonies).
3. Bourlière, F. (1963). Specific feeding habits of African Carnivores. *Afr. Wild Life*, *17*: 21–27.
4. Brehm, Alfred Edmund (1929). *Tierleben: Allgemeine Kunde des Tierreichs*. Leipzig.
5. Brown, L. H. (1963). Observations on East African Birds of Prey. *E. Afr. Wildl. J.*, *1*: 5–17.
6. Brown, W. L. (1960). Ants, acacias and browsing mammals. *Ecology*, *41*: 587–592.
7. Buechner, Helmut K. (1961). Territorial behaviour in the Uganda Kob. *Science*, *133*: 698–699.
8. Cott, Hugh B. (1940). *Adaptive Coloration in Animals*. London (Methuen).
9. Cott, Hugh B. (1946). The edibility of birds: illustrated by five years' experiments and observations on the food preferences of the hornet, cat and man. *Proc. zool. Soc. London*, *116*: 371–524.
10. Cott, Hugh B. (1948). Camouflage. *Advancement of Science*, *4* (16): 300–309.
11. Cott, Hugh B. (1954). Allaesthetic selection and its evolutionary aspects. *Evolution as a Process*, Ed. Julian Huxley. London (George Allen and Unwin).
12. Cott, Hugh B. (1956). *Zoological Photography in Practice*. London (Fountain Press).
13. Cott, Hugh B. (1959). *Uganda in Black and White*. London (Macmillan).
14. Cott, Hugh B. (1961). Scientific results of an inquiry into the ecology and economic status of the Nile Crocodile (*Crocodilus niloticus*) in Uganda and Northern Rhodesia. *Trans. zool. Soc. London*, *29*: 211–358.
15. Cott, Hugh B. and Benson, C. W. (1970). The palatability of birds, mainly based upon observations of a tasting panel in Zambia. *Ostrich. Sup. 8*: 357–384.
16. Cowles, Raymond B. (1959). *Zulu Journal*. Berkeley (Univ. Cal. Press).
17. Cullen, Anthony (1959). *Downey's Africa*. London (Cassell).
18. Darling, F. Fraser (1960). An ecological reconnaissance of the Mara plains in Kenya Colony. *Wildlife Monographs*, *5*: 1–41.
19. Darling, F. Fraser (1960). *Wild Life in an African Territory*. London (Oxford University Press).
20. Dice, Lee R. (1945). Minimum intensities of illumination under which owls can find dead prey by sight. *Amer. Nat.*, *79*: 385–416.
21. Dorst, Jean and Dandelot, Pierre (1970). *A Field Guide to the Larger Mammals of Africa*. London (Collins).
22. Elton, Charles (1927). *Animal Ecology*. London (Sidgwick & Jackson).
23. Estes, Richard D. (1967). The comparative behaviour of Grant's and Thomson's Gazelles. *J. Mamm.*, *48*: 189–209.

REFERENCES

24. Estes, Richard D. and Goddard, John (1967). Prey selection and hunting behaviour of the African Wild Dog. *J. Wildl. Management*, *31*: 52–70.
25. Field, C. R. (1970). Observations on the food habits of tame Warthog and antelope in Uganda. *E. Afr. Wildl. J.*, *8*: 1–17.
26. Fosbrooke, Henry (1972). *Ngorongoro, the Eighth Wonder*. London (André Deutsch).
27. Foster, J. B. (1966). The Giraffe of Nairobi National Park: home range, sex ratios, the herd, and food. *E. Afr. Wildl. J.*, *4*: 139–148.
28. Foster, J. B. and Dagg, A. I. (1972). Notes on the biology of the Giraffe. *E. Afr. Wildl. J.*, *10*: 1–16.
29. Friedmann, H. (1955). The Honey-guides. *U.S. Nat. Mus. Bull.* *208*: 1-292.
30. Goddard, John (1967). Home range, behaviour, and recruitment rates of two Black Rhinoceros populations. *E. Afr. Wildl. J.*, *5*: 133–150.
31. Graham, R. R. (1934). The silent flight of owls. *J. Roy. Aero. Soc.*, *38*: 837–843.
32. Grassé, Pierre-P. (1950). *Oiseaux. Traité de Zoologie*, *15*. Paris.
33. Guggisberg, C. A. W. (1969). *Giraffes*. London (Arthur Baker).
34. Howse, P. E. (1970). *Termites*. London (Hutchinson Univ. Libr.).
35. Innes, Anne C. (1958). The behaviour of the Giraffe, *Giraffa camelopardalis*, in the Eastern Transvaal. *Proc. zool. Soc. London*, *131*: 245–278.
36. Jackson, Frederick J. (1938). *The Birds of Kenya Colony and the Uganda Protectorate*. London.
37. Kemp, A. C. (1969). The general ecology of the Hornbills in the Kruger National Park. *Bokmakierie* (Sup.) *21* (3): 15–18.
38. Kruuk, H. (1966). Clan-system and feeding habits of Spotted Hyaenas (*Crocuta crocuta* Erxleben). *Nature*, *209*: 1257–1258.
39. Kruuk, H. and Sands, W. A. (1972). The Aardwolf (*Proteles cristatus* Sparrman) as predator on termites. *E. Afr. Wildl. J.*, *10*: 211–227.
40. Kruuk, H. and Turner, M. (1967). Comparative notes on predation by Lion, Leopard, Cheetah and Wild Dog in the Serengeti area, East Africa. *Mammalia*, *31*: 1–27.
41. Kühme, Wolfdietrich (1965). Communal food distribution and division of labour in African Hunting Dogs. *Nature*, *205*: 443–444.
42. Lamborn, W. A. (1913). Notes on habits of certain reptiles in the Lagos District. *Proc. zool. Soc. London*, *1913*: 218–224.
43. Lamprey, H. F. (1963). Ecological separation of the large mammal species in the Tarangire Game Reserve, Tanganyika. *E. Afr. Wildl. J*, 63–92.
44 Lawick–Goodall, Hugo and Jane van. (1970) *Innocent Killers*. London (Collins).
45. Laws, R. M. and Clough, G. (1965). Observations on reproduction in the Hippopotamus, *Hippopotamus amphibius* Linn. *Symposia Zool. Soc. London*, No. *15*: 117–140.
46. Laws, R. M. and Parker, I. S. C. (1968). Recent studies on Elephant populations in East Africa. *Symp. zool. Soc. London*, *21*: 319–359.
47. Leuthold, Walter (1970). Preliminary observations on food habits of Gerenuk in Tsavo National Park, Kenya. *E. Afr. Wildl. J.*, *8*: 73–84.
48. Mackworth–Praed, C. W. and Grant, C. H. B. (1952). *Birds of Eastern and North-eastern Africa*. London.
49. Maclaren, P. I. R. (1950). Bird-ant nesting associations. *Ibis*, *92*: 564–566.
50. Makacha, Stephen and Schaller, George B. (1969). Observations on Lions in the Lake Manyara National Park, Tanzania. *E. Afr. Wildl. J.*, *7*: 99–103.

51. Marshall, Guy A. K. and Poulton, E. B. (1902). Five years' observations and experiments (1896-1901) on the bionomics of South African insects, chiefly directed to the investigation of mimicry and warning colours. *Trans. ent. Soc. London*: 287-584.

52. Matthews, L. Harrison (1939). The bionomics of the Spotted Hyaena *Crocuta crocuta* Erxleben. *Proc. zool. Soc. London* (1939): 43-56.

53. Meinertzhagen, R. (1959). *Pirates and Predators*. Edinburgh (Oliver and Boyd).

54. Modha, M. L. (1967). The ecology of the Nile Crocodile (*Crocodylus niloticus* Laurenti) on Central Island, Lake Rudolf. *E. Afr. Wildl. J.*, 5: 74-95.

55. Moreau, R. E. (1936). The breeding biology of certain East African Hornbills (Bucerotidae). *J. E. Afr. and Uganda Nat. Hist. Soc.*, 43: 1-28.

56. Moreau, R. E. (1937). The comparative breeding biology of the African Hornbills (Bucerotidae). *Proc. zool. Soc. London*, 107A: 331-346.

57. Murphy, Robert Cushman (1955). Feeding habits of the Everglade Kite (Rostrhamus sociabilis). *Auk*, 72: 204-205.

58. Myers, J. G. (1935). Nesting association of birds and social insects. *Trans. Roy. Ent. Soc. London*, 83: 11-22.

59. Neal, Ernest (1970). The Banded Mongoose, *Mungos mungo* Gmelin. *E. Afr. Wildl. J.*, 8: 63-71.

60. North, M. E. W. (1942). The nesting of some Kenya Colony Hornbills. *Ibis* (14) 6: 499-508.

61. Owen, T. R. H. (1960). *Hunting Big Game with Gun and Camera in Africa*. London (Herbert Jenkins).

62. Pooley, A. C. (1969). Preliminary studies on the breeding of the Nile Crocodile *Crocodylus niloticus*, in Zululand. *Lammergeyer*, 3 (10): 22-44.

63. Poppleton, F. (1957). The birth of an Elephant. *Oryx*, 4: 180-181.

64. Pumphrey, R. J. (1948). The sense organs of birds. *Ibis*, 90: 171-199.

65. Rensch, B. (1957). The intelligence of Elephants. *Sci. Amer.* 196 (2): 44-49.

66. Roosevelt, Th. and Heller, E. (1922). *Life Histories of African Game Animals*. London.

67. Schaller, George B. (1968). Hunting behaviour of the Cheetah in the Serengeti National Park, Tanzania. *E. Afr. Wildl. J.*, 6: 95-100.

68. Schenkel, Rudolf (1966). On sociology and behaviour in Impala (*Aepyceros melampus* Lichtenstein). *E. Afr. Wildl. J.*, 4: 99-114.

69. Schmidt, K. P. (1919). Contributions to the herpetology of the Belgian Congo based on the collection of the American Museum Congo Expedition, 1909-1915. *Bull. Amer. Mus. Nat. Hist.*, 39: 385-624.

70. Shortridge, G. C. (1934). *The Mammals of South West Africa*. London.

71. Simon, Noel (1962). *Between the Sunlight and the Thunder*. London (Collins).

72. Spinage, C. A. (1968). Horns and other bony structures of the skull of the Giraffe, and their functional significance. *E. Afr. Wildl. J.*, 53-61.

73. Spinage, C. A. (1969). Quantative assessment of ectoparasites. *E. Afr. Wildl. J.*, 7: 169-171.

74. Stevenson-Hamilton, J. (1947). *Wild Life in South Africa*. London (Cassell).

75. Swynnerton, C. F. M. (1916). On the coloration of the mouths and eggs of birds. I. The mouths of birds. *Ibis* (10) 4: 264-294.

76. Sykes, Sylvia K. (1964). The Ratel or Honey Badger. *Afr. Wild Life*, 18: 29-37.

77. Temple-Perkins, E. A. (1955). *Kingdom of the Elephant*. London (Melrose).

78. Thorpe, W. H. and Griffin, D. R. (1962). The lack of ultrasonic components in the flight noise of owls compared with other birds. *Ibis*, *104*: 256–257.

79 Vesey-FitzGerald, D F. (1954). Wild life in Kenya. *Oryx*, *2*: 286–293.

80. Voeltzkow, A. (1899). Biologie und Enturichlung der äusseren Körperform von *Crocodilus madagascariensis* Grand. *Abh. Sench. Naturg. Gesell. 26*: 1–150.

81. Walther, Fritz R. (1969). Flight behaviour and avoidance of predators in Thomson's Gazelle (*Gazella thomsoni* Guenther 1884). *Behaviour, 34:* 184–221.

82. Watson, R. M. (1967). The population ecology of the Wildebeeste (*Connochaetes taurinus albojubatus* Thomas) in the Serengeti. A dissertation submitted for the degree of Doctor of Philosophy in the University of Cambridge. January, 1967.

83. Welty, Joel Carl (1962). *The Life of Birds*. Philadelphia and London (W. B. Saunders).

84. Williams, John G. (1954). The Quelea threat to Africa's grain crops. *E. Afr. Agric. J., 19* (3): 133–136.

85. Williams, John G. (1963). *A Field Guide to the Birds of East and Central Africa*. London (Collins).

86. Williams, John G. (1967). *A Field Guide to the National Parks of East Africa*. London (Collins).

Index

Figures in *italic* refer to pages on which drawings occur

Born in Leicestershire in 1900, Hugh Cott
was educated at Rugby, Sandhurst and
Selwyn College, Cambridge. He was a
soldier in Ireland in 1919-21, and during the
second world war in the Middle East as
an expert in camouflage. After teaching
education at Bristol University he was
Lecturer in Zoology first at Glasgow, then
at Cambridge, becoming Dean of Selwyn
College in 1966. Throughout his academic
career Dr Cott has also been a noted
explorer — to south-eastern Brazil, the Lower
Amazon and Marajó Island between 1923
and 1926, followed by zoological expeditions
to the Canary Islands and to many parts of
East and Central Africa. In recent years
he has regularly been a lecturer on Swan's
Big Game and Bird Safaris to East Africa.
He has published three previous books
and numerous scientific papers on a variety
of subjects, but is perhaps most widely
known for his work on the Nile Crocodile.
He is a founder member of the Society of
Wildlife Artists and a Fellow of the
Royal Photographic Society.

ISBN 0 00 219093 1